High-throughput screening and evaluation of combinatorial cell penetrating peptoid libraries

to identify organelle- and organ-specific drug delivery molecules

D1743580

High-throughput screening and evaluation of combinatorial cell penetrating peptoid libraries to identify organelle- and organ-specific drug delivery molecules

Zur Erlangung des akademischen Grades eines

DOKTORS DER NATURWISSENSCHAFTEN

(Dr. rer. nat.)

von der KIT-Fakultät für Chemie und Biowissenschaften

des Karlsruher Instituts für Technologie (KIT)

genehmigte

DISSERTATION

von

M.Sc. Ilona Diana Majella Wehl

aus Berlin

1. Referentin: Prof. Dr. Ute Schepers
2. Referent: Prof. Dr. Hans-Achim Wagenknecht

Tag der mündlichen Prüfung: 04.02.2019

Bibliografische Information der Deutschen Nationalbibliothek

Die Deutsche Nationalbibliothek verzeichnet diese Publikation in der

Deutschen Nationalbibliografie; detaillierte bibliographische Daten sind im Internet

über http://dnb.d-nb.de abrufbar.

1. Aufl. - Göttingen: Cuvillier, 2019

Zugl.: Karlsruhe (KIT), Univ., Diss., 2019

© CUVILLIER VERLAG, Göttingen 2019

Nonnenstieg 8, 37075 Göttingen

Telefon: 0551-54724-0

Telefax: 0551-54724-21

www.cuvillier.de

ISBN 978-3-7369-9996-1

eISBN 978-3-7369-8996-2

Die vorliegende Arbeit wurde unter Anleitung von Prof. Dr. Ute Schepers in der Zeit von März 2016 bis Januar 2019 am Institut für Toxikologie und Genetik (ITG) des Karlsruher Instituts für Technologie (KIT) und, in der Zeit von August 2017 bis Dezember 2017, in der Molecular Foundry am Lawrence Berkeley National Laboratory (LBNL) durchgeführt.

Table of contents

Zusammenfassung

Der globale pharmazeutische Arzneimittelmarkt wächst rapide. Für viele Krankheiten, wie zum Beispiel neurologische Erkrankungen, sind die Behandlungsmöglichkeiten jedoch nach wie vor schlecht. Therapeutika wie Proteine, DNA oder RNA mit vielversprechenden *in vitro* Ergebnissen scheitern in klinischen Studien an unzureichender Bioverfügbarkeit oder Aktivitätsverlust. Um dieses Problem zu bewältigen, können Wirkstofftransport-Systeme die pharmakokinetischen und pharmakodynamischen Eigenschaften verschiedener Therapeutika erhöhen. Diese Systeme können eine kontrollierte Freisetzung gewährleisten und die Spezifität von Medikamenten für das Zielgewebe verbessern und somit Nebenwirkungen auf das gesunde, umliegende Gewebe verringern. Seit einigen Jahren werden zellpenetrierende Peptide für den Wirkstofftransport untersucht, welche viele geeignete Eigenschaften aufweisen, jedoch auch Nachteile wie den schnellen Abbau durch Enzyme. *N*-substituierte Polyglycine, sogenannte Peptoide, imitieren die Struktur von Peptiden und behalten dabei die Fähigkeit zur Zellpenetration. Darüber hinaus zeigen sie im Vergleich zu Peptiden eine verbesserte Stabilität und Zellaufnahme. In dieser Arbeit wurden mehrere fluoreszenzmarkierte Peptoidbibliotheken synthetisiert und auf Zellaufnahme, intrazelluläre Lokalisation und Zytotoxizität *in vitro* und *in vivo* untersucht. Zur Aufklärung einer Struktur-Funktionsbeziehung und Identifizierung von Organell- und Organspezifität, wurden Peptoide, welche sich in der Länge des Rückgrats, der Zusammensetzung der Seitenketten, der Hydrophilie und der Fluoreszenzmarkierung, sowie zyklische Peptoide, untersucht. Ein hocheffizienter, automatisierter Evaluierungsansatz wurde verwendet, um eine neue Klasse von Peptoiden zu identifizieren, welche in die Mitochondrien eindringen. Diese Peptoide bestehen aus lipophilen und aromatischen Seitenketten und können hinsichtlich ihrer Zytotoxizität im niedrigen mikromolaren Bereich unterteilt werden. Während Peptoide mit mäßiger Lipophilie geringe zytotoxische Effekte aufweisen und sich als mitochondrienspezifische Transportermoleküle eignen, zeigen stark lipophile Peptoide eine selektive Toxizität für Krebszellen. Zytotoxische Peptoide wurden als mitochondrienspezifische Anti-Krebsmittel analysiert, mit vielversprechenden Ergebnissen in zwei- und dreidimensionaler Zellkultur. Im Hinblick auf den Transport von Medikamenten konnte die Zyklisierung von Peptoiden ihre Zytotoxizität verringern und zu Unterschieden in der Zellaufnahme, im Vergleich zu ihren linearen Gegenstücken führen. Während die

Zusammensetzung der Peptoid-Seitenketten einen starken Einfluss auf die intrazelluläre Lokalisation hatte, zeigte der Fluoreszenzfarbstoff wenig Einfluss auf die Zellaufnahme und die Organellspezifität des Peptoids. Um die Komplexität zu erhöhen, wurde die Organspezifität der Peptoide *in vivo*-Studien mit dem Zebrafisch als Modellorganismus analysiert. Die phänotypische Charakterisierung ergab verschiedene organspezifische Kategorien. Die interessanteste Kategorie war dabei die Kolokalisierung mit den olfaktorischen Neuronen, welche die Möglichkeit zum hirnspezifischen Transport ergab. Die Mikroinjektion eines olfaktorisch-spezifischen Peptoids, reich an aromatischen und unpolaren Seitenketten, in den Blutkreislauf von Zebrafischembryonen zeigte eine hohe Affinität des Peptoids zum Hirngewebe. Dadurch wurden Leitstrukturen für hirnspezifische Transporter identifiziert, die eine weitere Optimierung und Bewertung ermöglichen. Darüber hinaus wurden fluoreszierende polymere Nanopartikel auf zelluläre Aufnahme und hirnspezifische Anreicherung untersucht. Hohe Wasserlöslichkeit, gute Zellpenetrationsfähigkeit und geringe Toxizität zeigten ihre Biokompatibilität. Sowohl Nanopartikel als auch zellpenetrierende Peptoide haben ein großes Potenzial als Transporter für die organell- und organspezifische Wirkstoffabgabe, insbesondere im Hinblick auf das Gehirn als Zielgewebe.

Abstract

The global pharmaceutical drug market is rapidly expanding. However, for many diseases, such as neurological disorders, the treatment options are still poor. Therapeutics including proteins, DNA or RNA with promising *in vitro* results, fail in clinical trials due to insufficient bioavailability or loss of activity. To overcome this issue drug delivery systems can increase the pharmacokinetic and pharmacodynamic properties of various therapeutic agents. These systems can ensure a controlled release, improve the specificity of drugs to the target tissue, and therefore decrease side effects to the healthy surrounding tissue. Cell penetrating peptides have been investigated for drug delivery for several years, displaying many suitable properties but also disadvantages, such as fast degradation by enzymes. Poly-*N*-substituted glycines, so-called peptoids, mimic the structure of peptides by maintaining the ability to penetrate cells. In addition, they show improved stability and cellular uptake compared to peptides. In this work, several fluorescently labeled peptoid libraries were synthesized and investigated for cellular uptake, intracellular localization and cytotoxicity *in vitro* and *in vivo*. In order to elucidate a structure-function relationship and identification of organelle and organ targeting peptoids, differing in backbone length, side chain composition, hydrophilicity and fluorophore labeling, as well as cyclic peptoids, were investigated. A highly efficient automated evaluation approach was used to identify a new class of mitochondria penetrating peptoids. These peptoids are composed of lipophilic and aromatic side chains and can be subdivided, concerning their cytotoxicity in the low micromolar range. While peptoids with moderate lipophilicity display low cytotoxic effects and are suitable mitochondria specific transporter molecules, highly lipophilic peptoids show selective toxicity to cancer cells. Cytotoxic peptoids were analyzed as mitochondria-targeting anticancer agents with promising results in two-dimensional and three-dimensional cell culture. With regard to drug delivery, cyclization of peptoids could decrease their cytotoxicity and lead to differences in cellular uptake, compared to their linear counterparts. While the composition of the peptoid side chains had a strong impact on the intracellular localization, the fluorescent dye displayed little impact on cellular uptake and organelle specificity of the peptoid. To increase the complexity, the organ targeting of the peptoids was analyzed in *in vivo* studies by using the zebrafish as model organism. Phenotypic characterization revealed different organ-specific categories. The most interesting category was the colocalization with the olfactory neurons, which gave rise to the possibility for brain specific transport. Microinjection of an olfactory specific peptoid, rich in aromatic and nonpolar side chains, in the blood circulation of zebrafish embryos proved a high affinity of the peptoid to the brain tissue. Therefore, lead structures for brain specific transporters were identified, which allow further optimization and evaluation. Furthermore, fluorescent polymeric nanoparticles were investigated for cellular uptake and brain specific accumulation. High water-solubility, good cell penetration ability and low toxicity exhibited their biocompatibility. Both, nanoparticles and cell penetrating peptoids, have great potential as transporters for organelle- and organ-specific drug delivery, especially with respect to the brain as target tissue.

1. Introduction

1.1. Drug delivery

Over the past decades, the search for pharmacologically active compounds is growing and a tremendous progress in the treatment of various diseases has been made. A wide variety of small molecules, with a molecular weight less than 500 g/mol, peptides, proteins, nuclei acids and macromolecules were investigated. However, many active molecules with promising *in vitro* results fail in *in vivo* studies or clinical trials. Instead of specific accumulation of the drug in the desired tissue, they are evenly distributed in the body, leading to various side effects, e.g. cytotoxicity of drugs to the healthy surrounding tissue [1]. Adverse drug events lead to extended hospital stays and increased risk of mortality of patients and therefore to billions of health care costs per year [2]. Another major problem is the loss of *in vivo* activity, as biological barriers, such as cell membranes or tissue, are obstacles for therapeutic agents to reach their target. Furthermore, many drugs display poor pharmacokinetic properties, e.g. due to clearance by the kidney, fast enzymatic degradation of drugs, like peptides or proteins, or poor efficiency at physiological pH. Hence, most molecules with short half-lives are not suitable for *in vivo* applications or high invasive doses and continuous infusions, increasing the possibility for complications, are needed for successful treatment [3]. A further issue is the treatment of diseases of the central nervous system (CNS), such as Alzheimer's, Parkinson's disease or brain tumors, which are frequently occurring but their diagnosis is usually poor. Even though great effort in screening of potential drugs for CNS diseases is made the success rate is low and the global market for those drugs is very small. For most drugs overcoming the blood-brain barrier (BBB) is challenging, as only few, small and lipophilic molecules are able to cross this barrier [4-6]. The BBB is a selective, natural and essential barrier, separating the CNS from the peripheral blood [7]. Due to this boarder the brain is protected from circulating pathogens and thus brain infections are rare. The BBB is defined by an extremely tight blood vessel system in the brain. In contrast to normal capillaries, brain capillaries contain a tight endothelial cell layer surrounded by pericytes and astrocytes. These cells ensure the upregulation of tight junctions, neurons and a strong basal membrane [8-10]. Only lipophilic and small molecules can penetrate the membranes of endothelia cells and cross the barrier, while hydrophilic and charged molecules display no affinity to the lipophilic lipid membranes and are not able to cross membranes. In addition, even molecules which are predicted to cross the BBB can be discharged by efflux transporters, e.g. P-glycoprotein, which are the gate keepers of the BBB by serving as a transport barrier. Presently, treatments of CNS diseases often include risky invasive surgery or local application of drugs. Thus, possibilities for enhanced delivery of drugs to the brain, giving them the capacity to cross the BBB, are critical.

In general, the applicability of a pharmacological active drug can be calculated by dint of the therapeutic index (TI). The TI is defined by the proportion of lethal dose (LD_{50}), causing increased toxicity or death, compared to effective dose (ED_{50}), inducing the desired therapeutic effect [11, 12]. As the TI of drugs is often low, new methods for enhanced delivery are strongly needed, decreasing their toxicity and improving the effectivity. Thus, not only the development of drugs is a growing research field but also drug targeting. Specific accumulation in the target tissue can be achieved with passive and active drug targeting. Passive targeting benefits of abnormal properties of the target tissue e.g. wide fenestrations of blood vessels in tumors. Enhanced permeability and retention (EPR) effect leads to increased uptake of macromolecules and nanoparticles in the tumor tissue [13]. Active targeting can be achieved with cell- or organ- specific ligands. For both, passive and active targeting, drug delivery systems (DDS) can be used. DDS enhance cellular uptake, increase the solubility of drugs and protect them from degradation by enzymes. Cargos can be transported across various biological barriers, even the BBB, to specific cellular organelles or target organs. Therefore, side effects to the healthy surrounding tissue can be reduced and similarly an improvement of dose-efficiency relationship is possible. Hence, the pharmacological properties and the therapeutic indices of drugs can be strongly improved.

1.2.   Drug delivery systems

Enhanced cellular uptake of drugs or organ-specific delivery and controlled release can be achieved by attachment of drugs to various drug delivery systems (DDS). Over the last decades a wide range of DDS have been investigated for biocompatibility, cell penetration abilities, pharmacokinetic properties, controlled drug release and ability to transport cargos. Important properties for DDS are low toxicity, intense stability *in vivo*, with avoidance of long-term accumulation in the body, and the ability to cross insistent barriers, e.g. the BBB [14]. Carrier molecules can be for example nanoparticles, liposomes, antibodies, polymers, viruses, micelles or peptides. Each DDS has different advantages and disadvantages and therefore not all carrier systems might be suitable for all drugs or target organs. Liposomes for example are

6

prone to encapsulate drugs, to increase their solubility and decrease their cytotoxicity [15]. However, encapsulation is not suitable for neutral hydrophobic molecules as they are rapidly released [16]. Nanoparticles are especially interesting in the field of cancer therapy as they show increased accumulation in tumor tissue. Unfortunately, many nanoparticles display increased cytotoxicity. Cell penetrating peptides have been emphasised by the literature as they have the ability to cross different cell membranes and display low toxicity *in vivo* and *in vitro*.

1.3. Cell penetrating peptides

In 1988 Green and Loewenstein, as well as Frankel and Pabo, discovered that the trans-activator of transcription (Tat) protein of HIV-1 is able to cross the membrane of various cells and could even reach the nucleus [17, 18]. At that point in time this was a completely new finding, as only one year before Sternson et al. reported that proteins and peptides display poor cellular uptake and might therefore not be suitable as drug candidates [19]. The Tat protein consists of 86 amino acids and it could be shown that it is rapidly taken up by cells and stimulates HIV-LTR driven RNA synthesis [17]. Green and Loewenstein were also able to synthesis active mutant peptides, with 21 to 41 amino acids. In the following years the Tat protein was extensively analyzed and several mutant peptides have been investigated to identify the minimal sequence required for cellular uptake. Vivès *et al.* discovered in 1997, that the basic domain Tat-(48-60) is needed for cellular uptake and deletion or substitution of amino acids within this domain reduced the cellular uptake. This cationic area consists mainly of arginine and lysine side chains. Further investigations, to find the shortest structure needed for cellular uptake, showed that the peptide can be reduced to an eleven amino acids motif, Tat-(47-57), which is still able to translocate cells. In 1991 it was discovered, that the 60 amino acid polypeptide of the Drosophila Antennapedia homeodomain (pAntp) was able to penetrate cells as well [20]. Based on these results, in 1994, a 16-mer cell penetrating peptide, so called penetratin, derived from the third helix of Antennapedia was synthesized. As internalization of the peptide was also found at 4 °C, it was assumed that the peptide is taken up by an energy-independent mechanism [21]. In the following years many further cell-penetrating peptides, derived from protein transduction domains, were found, such as transportan, a synthetic peptide built from the neuropeptide galanin linked to mastoparan, or VP22, based on a sequence of herpes virus-type 1 virus [22, 23]. Additionally, it was discovered that not only peptides rich in cationic amino acids are able to penetrate cells but also amphipathic peptides, consisting of a hydrophilic and a hydrophobic domain. Therefore, amphipathic peptides contain not only lysine and arginine but also hydrophobic amino acids like alanine, leucine, isoleucine and valine [24]. The class of amphipathic cell penetrating peptides can be subdivided in three groups: primary, secondary and proline-rich peptides. Primary amphipathic peptides are derived from a sequential assembly of hydrophobic and hydrophilic domains in their primary sequence, for example MPG or Pep-1 [25]. In contrast, secondary amphipathic peptides are received after folding in secondary structures, mostly α-helical structure, leading to hydrophobic domains opposite to hydrophilic domains, e.g. peptides in the MAP (model amphipathic peptide) family [26]. Last, but not least, hydrophobic peptides, containing mainly nonpolar amino acids, were found to have a high affinity to hydrophobic cellular membranes, giving them the ability to translocate the membrane spontaneously. As an example, C105Y, derived from α-1-antitrypsin, can be mentioned [27]. However, compared to cationic and amphipathic peptides, the amount of discovered hydrophobic peptides is low and therefore knowledge about them is small [28]. Table 1 gives an overview of different, well characterized cationic, amphipathic and hydrophobic cell penetrating peptides.

Table 1: Overview of eight well-studied cationic, amphipathic and hydrophobic CPPs. Sequences are shown in one letter code for each CPP [28]

Peptide	Sequence	Class
Tat-(47-57)	YGRKKRRQRRR	Cationic
Penetratin	RQIKIWFQNRRMKWKK	Cationic
Transportan	GWTLNSAGYLLGKINLKALAALAKKIL	Cationic
VP22	DAATATRGRSAASRPTERPRAPAR-SASRPRRPVD	Cationic
MPG	GALFLGFLGAAGSTMGAWSQPKKKRKV	Amphipathic
Pep-1	KETWWETWWTEWSQPKKKRKV	Amphipathic
MAP	KALAKALAKALA	Amphipathic
C105Y	CSIPPEVKFNKPFVYLI	Hydrophobic

As CPPs are highly charged, mostly due to cationic side chains, and accumulate not only in endosomes but also in cytoplasm and the cell nucleus, toxicity of CPPs was intensively studied. Low impacts on the viability of adherent and nonadherent cell lines, in low micromolar range, was found for most CPPs. Only pAntp displays increased toxicity for concentrations above 50 μM, whereas it was shown for Tat(48-57) that even concentrations of 100 μM were harmless to various cell lines [29, 30]. These results were also confirmed in *in vivo* studies, finding no visible toxicity after injection of a fusion protein in mice [31].

1.3.1. Mechanisms of cellular uptake

So far, it is known that specific peptides are able to cross the membranes of several cell types, however the mechanism of cellular uptake of peptides has not been completely understood and it is assumed that different pathways are possible. There are probably various factors, such as peptide structure and size, cell type, temperature, concentration and incubation time, which play a crucial role for the route to enter the cell [32]. However, it can be differentiated between two entry mechanisms: the energy-dependent endocytosis and direct penetration of the membrane, which is energy-independent. Initially, it was believed that uptake is only taken place by direct penetration, however, these findings were artifacts, due to protocols involving fixation of cells [21, 33, 34]. Lately, it has been assumed, that the most frequent pathway to enter the cell for CPPs is endocytosis [35-37]. Possible endocytosis ways are phagocytosis, macropinocytosis, caveolin- or clathrin-mediated endocytosis and clathrin/caveolin-independent endocytosis [38]. Phagocytosis, an active pathway, which is possible only in specific cell types and involves several specific cell-surface receptors, usually takes place for clearance of large particles, such as pathogens or apoptotic cells and is therefore probably less important for CPPs [39]. More important for CPPs are pinocytosis pathways, occurring in all mammalian cell types. It was found, that CPPs are able to use more than one pathway simultaneously, for example for Tat marcopinocytosis, clathrin-mediated as well as caveolin-mediated endocytosis was found [37]. For direct penetration there are also different possible ways: formation of pores, formation of inverted micelles and the carpet model. It is reported, that amphipathic CPPs are able form pores in the membrane, for example due to their α-helical structure. Hydrophobic side chains in the helix facing the membrane, while hydrophilic side chains are hidden inside the pore [40]. The carpet model suggests that peptides self-associate and lie parallel on the membrane, interacting with their cationic side chains with the negative charges of the phospholipid headgroups on the membrane surface. Thus, membrane organization is disturbed and peptides are able to cross the membrane border [41]. Furthermore, the inverted micelle model was found, initially for pAntp. CPPs interacts with the membrane, leading to the formation of inverted micelles, encapsulating the peptides. Further interactions of CPPs with the membrane, in the inverted micelle, destabilizes the membrane and CPPs are released into the cell [21, 42, 43]. An overview of different possible pathways is shown in figure 1.

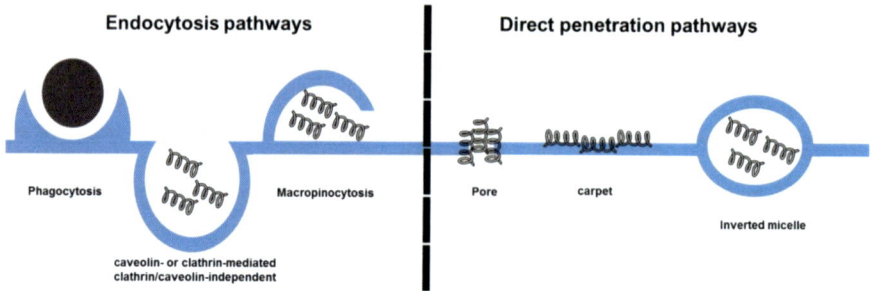

Figure 1: Different possible pathways for uptake of CPPs are shown. Left: three possible ways of endocytosis: phagocytosis, caveolin- or clatherin-mediated and caveolin/clatherin-independent pathway and micropinocytosis. Right: Three ways for direct penetration: formation of pores, carpet model and inverted micelles. Cell membrane is pictured by a blue barrier and CPPs are represented by short helices. Source: [28, 44]

1.3.2. Application of cell penetrating peptides

Since the discovery of cell penetrating abilities of peptides not only peptides themselves have been studied but also their capability to transport cargo into cells. For their application as transport vectors there is the possibility of covalent conjugation of the cargo to the peptide or complexation based on non-covalent interaction. Suitable covalent conjugation are for example amide bonds, disulfide bonds or thioester linkages as they give the possibility to release the cargo after penetration of the cell membrane [45-47]. Covalent binding of cargo has been shown to be suitable for the transport of various molecules, under reproducible conditions, but it might also inhibit the biological activity of the cargo [48]. Non-covalent conjugation can be achieved by simple mixing of cargo and CPP, due to hydrophobic or electrostatic interactions, depending on physiochemical properties of peptide and cargo. Non-covalent bindings are easier to achieve, however, resulting complexes are hard to control and variations in size and composition occur, whereas compositions for covalent bindings are defined [47]. In 1998 Nagahara *et al.* was able to conjugate Tat peptide covalently to proteins and could successfully show transport abilities of Tat *in vitro* [49]. In 2001, Morris *et al.* could demonstrate efficient delivery of peptides and proteins, non-covalently linked to Pep-1, into mammalian cells. Pep-1 was highly recommended, due to low toxicity and good stability in the presence of physiological buffers [50]. Since then, various cargo could be transported into cells, including macromolecules, DNA, RNA and proteins [51-53]. Furthermore, CPPs were investigated for delivery across the BBB, finding promising results, as it could be shown, that Tat is able to overcome the barrier and deliver Bcl-xl, which plays a role in neuronal apoptosis, in the brain tissue and protecting it against ischemic brain injury [54]. Tat was also fused to ß-galactosidase and after injection in mice ß-Gal activity was not only found in different organs, including kidney, heart and lung, but also in the brain [55]. Rouselle et al. compared doxorubicin (dox) concentrations, a well-known chemotherapeutic agent, of free dox to dox coupled to CPPs in rat brains. It could be shown that CPPs significantly increased the dox concentration in the brain [56]. To summit up, many promising results have been found for CPPs as delivery vectors, including low toxicity, strong cellular uptake and the ability to transport cargo into cells.

1.3.3. Limitations of CPPs

Despite the numerous advantages of CPPs they also have disadvantages, limiting their usage as suitable transporter vehicles for drug delivery. For most CPPs endosomal uptake is found, and even though they internalize cells in this manner, they are still encapsulated by endosomal membranes, unable to deliver their cargo in cytoplasm or to the nuclei [57]. Endosomal escape is hard to achieve, as it requires destabilization of endosomal membranes, accomplishable only by adding cytotoxic auxiliary compounds. Major drawbacks can also be found regarding their application *in vivo*. As they are able to penetrate different cell lines, most CPPs are found to be not cell or tissue specific. Therefore, CPPs lack of specificity for delivery of cargo *in vivo* and distribution is found in various parts of the body in animal models [55, 58]. After fusion of CPPs to antibodies, cellular uptake of antibodies could be improved significantly, however, also unspecific enrichments were found in different non-targeted tissues [59, 60]. The major problem of peptides, however, is their limited usage *in vivo*. Due to their natural and peptidic structure rapid degradation takes place in the presence of serum or intestinal fluids [61]. Thus, CPPs display poor pharmacokinetic properties and high doses are needed for therapeutic effects.

1.4.   Peptidomimetics

As the demand of peptides which are more stable *in vivo*, is growing, an important and promising research field are peptidomimetics. These non-natural molecules mimic natural peptides but due to modifications in the backbone or side chains they can exhibit increased stability against proteolysis. Due to modifications in the peptide structure improved cellular uptake and less nonspecific receptor binding can be achieved, leading to enhanced pharmacokinetic and pharmacodynamics properties [62]. Stability of peptides can already be improved by integration of non-natural side chains or changing the natural occurring L-amino acids to D-amino acids (*inverso*). Additionally, the order of amino acids can be reversed (retro), to keep stereochemistry to their peptide counterparts. Stability of *retro inverso* peptides is improved however, they also display increased toxicity to cells [63]. Another possible approach to increase pharmacological properties of peptides are for example ß- or γ-peptides, consisting of ß- or γ-amino acids, which have their amino group bound to ß- or γ-carbon instead of the α-carbon. Shifting of the amino group leads to an extension of the peptide backbone and due to increased conformational flexibility to changes in secondary structures of those peptides. For ß-peptides not only enhanced stability against several peptidases has been shown but also improved biological activity [64, 65]. In addition, it is possible to replace one or more amino groups in the backbone to an oxygen atom, so called depsipeptides. This change induces structural changes and more flexible structures, as hydrogen bonds are decreased after elimination of the amino group. Several depsipeptides have been synthesized or isolated from bacteria and studied for antimicrobial activity with promising results. Furthermore, cyclization of peptides can improve their bioavailability, as amide bonds are sheltered in the cycle, enhancing cellular uptake and decreasing enzymatic degradation. Structures of exemplary peptidomimetics are shown in figure 2.

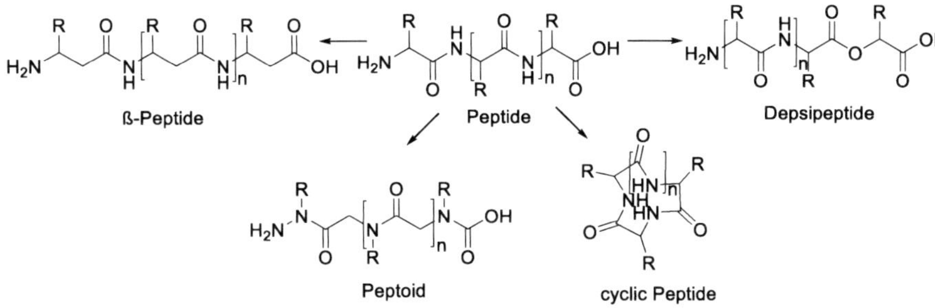

Figure 2: Structure of peptide backbone and four different peptidomimetics: ß-peptide, depsipeptide, peptoid and cyclic peptide. [66]

Another possible modification is shifting the side chain from the α-carbon to the nitrogen atom in the peptide backbone, receiving poly-N-substituted glycines, so called peptoids.

1.5. Peptoids

Peptoids are not only stable in presence of enzymes, they also display enhanced stability to chemicals or high temperature. Miller at el. compared stability of peptides and peptoids to proteolytic enzymes, such as trypsin, pepsin and papain, finding rapid hydrolysis of peptides and in contrast to that, stable and unchanged peptoids [67]. Comparing absorptive clearance, absorption and disposition of peptides and peptoids *in vivo* shows, that they are similar concerning their clearance, but differ in absorption and disposition [68]. Thus, peptoids are stable in the body, finding no metabolization, but elimination is fast, probably due to high lipophilicity. Rapid clearance can also be advantageous, as long-term tissue accumulation might cause cytotoxic effects. In contrast to many other peptidomimetics, peptoids display strong cellular uptake combined with low toxicity [69]. Peptoid synthesis is relatively easy and cost-efficient and gives the possibility to integrate a wide range of non-natural side chains. Peptoids with a large sequence diversity can be synthesized, as there are currently over 300 possible side chains available [70]. Due to the variation in the peptide backbone, secondary structures of peptoids differ to peptides as they have no free amine group in the backbone and therefore lack of a hydrogen-bond donor. However, by inserting specific side chains secondary structures, such as helices, which are stable to chemicals and temperature, can be received [71].

1.5.1. Application of peptoids

There is a wide field of application for peptoids in drug-discovery and -delivery and material science. Peptoids themselves can serve as drugs and have been investigated for antimicrobial activity, as counterparts to antimicrobial peptides. Lately, the amount of infections with pathogens, resistant to known antibiotics, is increasing, acquiring hospital stays or death of patients [72, 73]. The necessity for new antimicrobial substances is indispensable and promising results have been found for peptoid activity against several gram-positive and gram-negative bacteria, including Methicillin-resistant *Staphylococcus aureus,* fungi and parasites [72, 74-76]. Their antimicrobial activity is enhanced compared to peptides, due to decreased proteolytic degradation and in comparison to peptides they display no immunogenicity [77]. Even for short, dimeric and trimeric peptoids promising *in vitro* and *in vivo* results were found [78]. Furthermore, peptoids are investigated as anticancer agents, as alternative to commonly used chemotherapeutical agents, causing various side effects or incomplete killing of cancer cells, leading to the expression and enhanced activity of multidrug-resistant-transporters. Cytotoxic effects of peptoids to a wide spectrum of cancer cell lines, as well as multidrug resistant cells, could be shown, with significant lower toxicity to primary cells [77, 79]. Potent peptoids were found for breast, lung and prostate cancers showing specific *in vitro* results and significantly decrease of tumor growth in animal models [77, 79, 80]. In addition peptoids are able to bind specific to receptors or enzymes acting as inhibitors or activators [81]. Peptoid inhibitors exhibit strong affinity to enzymes and are able to compete with the natural substrate or their analog peptide inhibitor [82, 83]. Hamy *et al.* could discover a peptoid/peptide hybrid inhibiting interactions between Tat and TAR RNA, a crucial step in HIV gene expression [84]. In terms of material science, it has been shown that peptoid polymers with alternating charged and aromatic side chains are self-assembling, due to the hydrophobic effect, to ordered 2D nanosheets, after rocking them in polar media. Functionalized nanosheets can be used for specific binding of proteins, bacteria or metal ions [85-87].

Beyond that, peptoids are also suitable molecular transporters as counterparts to cell penetrating peptides. According to cell penetrating peptides, many cationic peptides, rich in arginine and lysine, have been studied for cell penetration abilities, toxicity and intracellular localization [69, 88-90]. Not only cationic peptides are able to penetrate cells but also amphipathic and highly lipophilic peptoids [88]. Due to the possibility to synthesis diverse peptoid libraries, including non-natural side chains, different cell organelles or *in vivo* specific organs can be addressed [91]. A broad spectrum of functional moieties can be introduced, giving peptoids several different properties, e.g. by copper-catalyzed alkyne azide cycloaddition (CuAAc), such as charged side chains, sugar moieties, nucleobases and fluorophores [92-94]. According to CPPs their mechanism of cellular uptake is not completely understood, including different endocytosis pathways and direct penetration of membranes [88, 89, 91, 95, 96]. Peptoids are suitable transfection agents for gene delivery in several cell lines, being even more effective compared to well-studied transfection lipids [97]. Additionally, short cell penetrating peptoids are even more cell permeable than the corresponding peptides, assuming that membrane penetration is enhanced, due to increased lipophilicity because of the lack of the polar amino bond [98]. Promising results have also been found in comparison of peptoids with the cell penetrating peptide Tat(49-57), as peptoids displayed better cellular uptake, resistance to proteases, low toxicity and synthesizes is easier and more economically [90].

1.5.2. Synthesis of peptoids

Equally to peptides, peptoids are synthesized *via* solid phase synthesis, developed 1963 by Merrifield [99]. Side chains are stepwise added to a growing peptide or peptoid chain, which is bound to a solid resin. In comparison to peptide synthesis in solution, good yields with high purity can be achieved, as reagents can be added in excess and separated, after completed reaction, by filtration. Hence, reagents can also be reused, making solid phase synthesis more economically. Another advantage for solid phase synthesis is the possibility of automated synthesis. Up to 30 compounds, in 0.1 mmol scale, or one-bead-one-compound libraries, containing up to thousand peptoids, can be fully automated synthesized in parallel [100]. The solid phase, used for peptoid synthesis, consists of small polystyrene beads, functionalized with a linker. Suitable resins for peptoid synthesis are Rink amide-, Sieber amide-, NovaSyn®TGR-resin or 2-chlorotritylchlorid resin. Amide functionalized resins are, with an exception of NovaSyn®TGR-resin, usually protected with 9H-Fluoren-9ylmethoxycarbonyl (Fmoc), a protection group, which can be removed by weak bases, such as 20% piperidine in Dimethylformamide (DMF). Cleavage of Fmoc-group is shown for Rink amide resin in scheme 1.

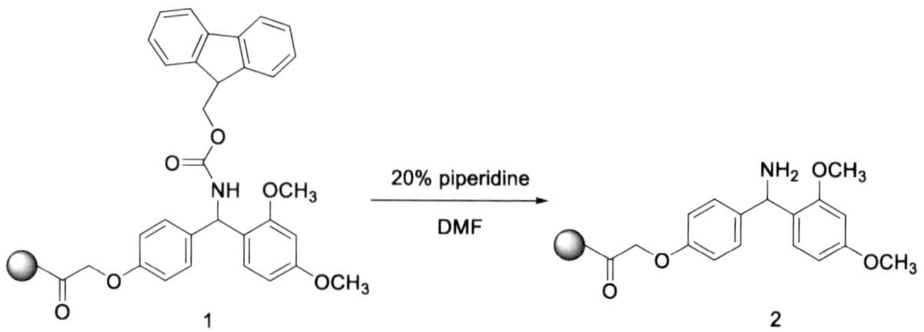

Scheme 1: Fmoc deprotection of the rink amide linker with 20% piperidine.

Deprotection can be controlled photometrically by UV/Vis spectroscopy, as the resulting fluorene is highly fluorescent (Ex.: 295 nm). For the following synthesis of the peptoid chain two methods are possible. Initially, peptoids were synthesized, equally to peptides, by monomeric method (scheme 2, A). N-substituted glycine monomers have to be protected with Fmoc and subsequently the can be coupled to the free secondary amine group on the resin, by activation with 1-hydroxybenzotriazol (HOBt) and N,N'-diisopropylcarboiimide (DIC). Afterwards Fmoc group is deprotected with 20% piperidine and the cycle can be repeated until desired peptoid length is obtained. Finally, the peptoid is cleaved from the resin with 95% trifluoroacetic acide (TFA). An advantage of this method is that Fmoc deprotection can be controlled after each coupling by measuring the UV absorption. However, synthesis of Fmoc protected N-substituted glycine monomers is very time-consuming and synthesis yields are low and therefore monomeric method is rather inefficient. In 1992 Zuckermann developed the submonomer method and efficiency, yields and side-chain diversity of peptoid could be increased, as this method does not involve Fmoc protection and deprotection steps. The secondary amine group on the resin has to be acylated with a haloacetic acide with DIC. Usually bromoacetic acide is used, as it was found to be more stable than iodoacetic acide and yields could be increased in comparison to chloroacetic acide [100]. Addition of 1-hydroxybenzotriazol (HOBt) is not needed and would even lower the synthesis yield. Afterwards the bromide can be replaced by nucleophilic substitution (S$_N$2 reaction) with a primary amine. The cycle can be repeated and final cleavage is also done with 95% TFA [100]. As the diversity of commercially available primary amines is large and synthesis yield are good, the submonomer method is suitable for the synthesis of large combinatorial peptoid libraries.

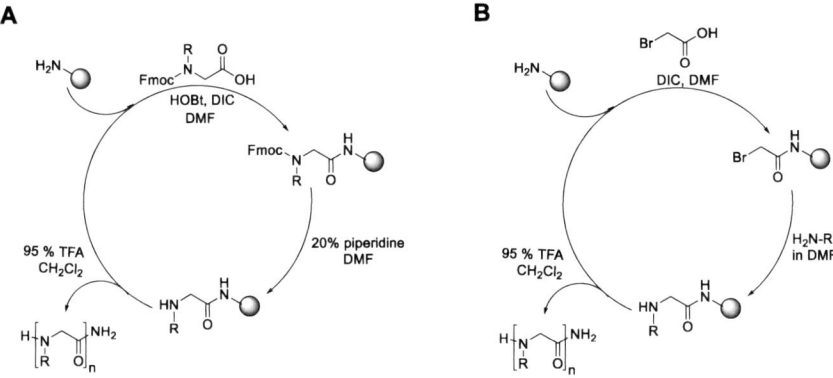

Scheme 2: Synthesis of peptoids by monomer method (A) and submonomer method (B).

1.5.3. Nomenclature of peptoids

So far, there is no universally used nomenclature for peptoid building blocks. All codes start with a cursive *N* at first position, indicating the *N*-substitution of the side chains, followed by a three-letter-code, according to amino acids. For monomer analogs to natural amino acids the standard three-letter amino acid abbreviation is used, e.g. *N*phe or *N*lys. For non-natural side chains the code in this work may differ to those found in the literature. For aromatic side chains position two indicates the position of the substituent, other than hydrogen, on the phenyl group (ortho, meta and para). The third and fourth position are an abbreviation for the functional group of the side chain. For aliphatic side chains the three-letter-code is an abbreviation of the monomer name. The nomination of peptoid chains is done from the *N*-terminus to the *C*-terminus. Structures, names and abbreviations of peptoid monomers which were used in this work are shown in table 2.

Table 2: Overview of submonomers used in this work. Structure, full name and abbreviation is shown for each building block.

Structure	Full name	Abbreviation
H_2N ~~~~ NH_2	Diaminobutan	Nlys
propargylamine structure NH_2	Propargylamine	Nprg
butylamine structure NH_2	Butylamine	Nbut
hexylamine structure NH_2	Hexylamine	Nhex
heptylamine structure NH_2	Heptylamine	Nhep
N_3 ~~~ NH_2	Azidopropylamine	N4az
benzylamine structure NH_2	Benzylamine	Nphe
HO — benzene — NH_2	4-Hydroxybenzylamine	Npob
F — benzene — NH_2	4-Fluorobenzylamine	Npbf
Cl — benzene — NH_2	4-Chlorobenzylamine	Npcb

1.5.4. Synthesis of peptoid libraries

Synthesis of large compound libraries is a useful method to identify the highest possible amount of positive lead structures in drug discovery and optimization. Combinatorial synthesis can be used for the synthesis of compounds libraries with the same basic structure but varying combinations or permutations of building blocks. The automated solid phase method could speed up the synthesis of peptides and peptoids, however, strategies are needed for the synthesis of hundreds of compounds at the same time. In 1984 Geysen at al. described "multipin" method for the simultaneous synthesis of large numbers of peptides and could successfully synthesize and analyse 208 hexapeptides. Peptides were synthesised on functionalized polyethylene rods (pins) and pins were assembled into a polyethylene holder with the format of a microtiter plate. Microtiter plates were prepared with reagents for solid phase peptide synthesis and unique reactions on individual pins could be performed at the same time. Efficiency of the synthesis could be improved, as substances could be divided from each other for individual coupling steps, but they could also be combined for washing steps. Subsequently peptides were tested for biological activity without removing them from the solid support [101]. Later, after further development of the method is was also possible to synthesis peptides in micromole scale [102]. Houghton et al. was also able to describe a method for the synthesis of peptides in micromole scale, by development of the "tea bag" approach. Resin beads were filled in small, labeled polypropylene mesh bags (15 x 20 mm). Due to the size of the tea bag mesh the resin cannot escape the bags but solvents and reagents are able to enter and react. A huge progress in the synthesis of libraries was made in 1991 when Furka *et al.* developed the spit-mix approach [103]. This method is highly efficient as it allows simultaneous synthesis of different compounds, varying in their composition of building blocks. For the initially split step, resin beads are divided in multiple equal portions, depending on the amount of building blocks. Afterwards different portions are coupled to the respective building block. In the following step beads are mixed again for washing steps, deprotection steps, or in the case of peptoids, synthesized by submonomer method, acylation steps. Subsequently, beads can be divided again and elongation of the peptide/peptoid chain can be done until the desired length is observed. In this manner large libraries can be synthesized with exponentially increasing numbers of compounds in each coupling step. The number of compounds in the library (n^m) can be calculated by the amount of building blocks (n) to the power of the length of the chain (m). For a library with three different building blocks three compounds are observed in the first cycle, nine in the second, 27 in the third and already 81 in the fourth. Split-mix approach for three different building blocks and two cycles is shown in figure 3.

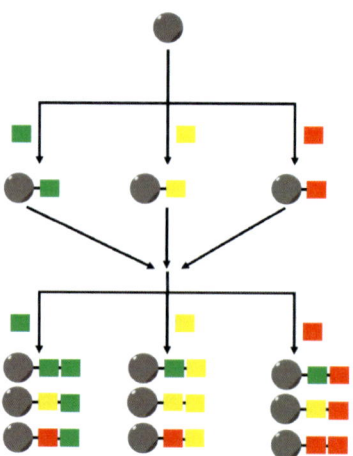

Figure 3: Schematic representation of the split-mix approach for a dimeric library containing three different building blocks. Solid resin beads are shown in grey and building blocks are represented in red, yellow and green.

Based on the split-mix approach, one bead-one compound (OBOC) synthesis is a powerful tool for synthesis and screening of large libraries. Libraries containing millions of compounds can be synthesized in a highly efficient manner, by random split-mix synthesis. After each mixing step beads are randomly dived for the coupling of different building blocks without following the chemical history of the beads. The amount of different compounds depends on the number of beads and building blocks and the same compound can occur several times in the library [104]. The advantage of those libraries is that each bead contains only one type of compound with up to 10^{13} copies on a 100 μm diameter bead [105]. Screening of OBOC libraries is usually done without cleavage of the compound from the bead. As sequences of individual beads are unknown, exact analysis is only done with "positive hits", displaying biological activity in the performed screening.

1.5.5. IRORI based synthesis

To investigate peptoids for their suitability as molecular transporters to different target tissues large, diverse compound libraries in millimolar scale, cleaved from the solid support, are needed. Furthermore, for structure-function relationships it is indispensable to know the exact structure of each compound in the library. These requirements cannot be achieved with one bead-one compound libraries, giving only small amounts of peptoids, or by time consuming "tea bag" synthesis, which does not offer the possibility for automated synthesis. The IRORI® technology, is a promising further developed method for large split-mix libraries with good yields [106-108]. Small polypropylene vessels, so called MiniKans, are filled with resin beads which have with optimal pore sizes, allowing solvents to enter, comparable to "tea bags". In contrast to the "tea bags", however, labelling is not done manually, but operated by adding a radiofrequency (Rf) tag in each Kan. Rf tags are small (13 mm x 3 mm) and stable in the presence of chemicals. Coding on Rf tags can be read with a USB device, giving not only the possibility of manual directed sorting of Kans but also fully automated sorting procedure with the IRORI AutoSort 10Kx instrument. The synthesis can be performed in standard glassware and libraries can be synthesized with a minimal volume of solvents and reagents. At any time, during the synthesis process, exact sequence of compounds in the MiniKans is known. Finally, cleavage of peptoids from the resin, followed by collection in multiwell plates can be done in a cleavage station, allowing simultaneous cleavage of up to 96 peptoids. Additionally, combination with LC/MS and MALDI-TOF-MS is possible, simplifying the control and tracking of the synthesis [106].

2. Objective

The aim of this work was the synthesis of several diverse peptoid libraries, differing for example in their composition of side chains, in IRORI MiniKans. Subsequently, high-throughput screenings should be performed *in vivo* and *in vitro*. Peptoids should be investigated for their biocompatibility, cellular uptake, organelle and organ specificity. The focus should be on brain specific peptoids, allowing the transport of drugs in the CNS. Peptoids with high and specific uptake should be further analyzed for their cytotoxicity to confirm their suitability as molecular transporters. A detailed variation of the peptoid sequence should give rise to a structure-function relationship for organ and organelle targeting. Finally, suitable transporter peptoids should be identified and further analyzed or modified. An overview of the evaluation process is shown in figure 4.

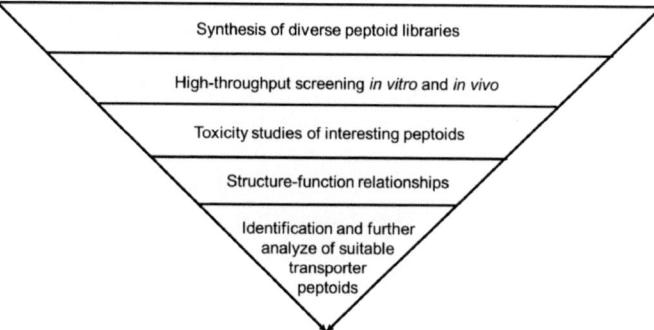

Figure 4: Overview of the process for analyzation of peptoid libraries and identification of suitable transporter peptoids for drug delivery.

The last part of this work is concerning the biocompatibility of polymeric nanoparticles and their usage as transporter vehicles. Therefore, they should be investigated for cellular uptake, intracellular localization, organ specificity and cytotoxicity.

3. Results and Discussion

3.1.   Molecular Transporters

3.1.1. Peptoid library 1

In a previous work, Franziska Rönicke (Institute of Toxicology and Genetics, KIT) investigated a tetrameric cell-penetrating peptoid library, containing a mixture of hydrophilic and lipophilic peptoids, to evaluate their organelle and organ specificity [91]. Peptoids of the library are consisting of *N*-4-aminobutylglycine (*N*lys, R₁), *N*-2-prop-2-yne-1-ylglycine (*N*prg, R₂), *N*-benzylglycine (*N*phe, R₃) and *N*-(p-chlorobenzyl)glycine (*N*pcb, R₄), sorted by increasing lipophilicity. Side chains are shown in figure 5 and are represented by different colors (*N*lys = blue, *N*prg = green, *N*phe = yellow, *N*pcb = red) to facilitate the presentation of the peptoid sequences in the following evaluations of the peptoid library.

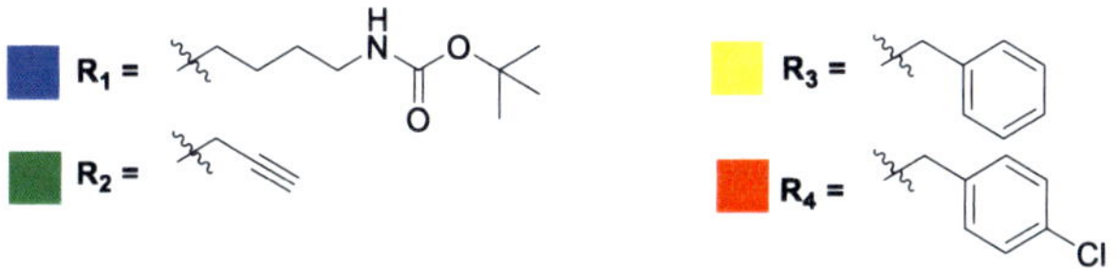

Figure 5: *N*-substituted glycine residues used for synthesis of peptoid library 1: tert-butyl (4-aminobutyl)carbamate (*N*lys, blue, R₁), prop-2-yne-1-amine (*N*prg, green, R₂), benzylamine (*N*phe, yellow, R₃) and 4-chlorobenzylamine (*N*pcb, red, R₄).

*N*phe and *N*lys represent two natural amino acids, phenylalanine and lysine. *N*-4-aminobutylglycine has two amine groups, so one had to be protected with a tert-butyloxycarbonyl (Boc) group, to prevent side reactions during the prolongation of the peptoid. Deprotection of the acid labile Boc group takes place during the cleavage from the resin (Scheme 1). *N*pcb side-chains were used to increase lipophilicity of peptoids and therefore increasing the variety of functionality of different lipophilic moieties. Additionally, the alkyne group of *N*prg residues allows for the coupling of azide building blocks by 1,3-dipolar cycloaddition and thiols by thiol-ene chemistry [109-111]. Hence, peptoids can be further modified to functionalize peptoids, incorporating additional groups, such as sugar moieties or cargos. The synthesis of the tetrameric peptoids was performed by submonomer solid phase method in combination with the split-mix approach using IRORI MiniKans. For each peptoid sequence of the library an appropriate Kan was filled with resin and a glass coated radio frequency (Rf) tag. Rf-tag codes can be read with an USB device, giving the advantage to

23

mix MiniKans for initial deprotection of the rink-amide resin with piperidine, and following acylation with bromoacetic acid and N,N'-diisopropylcarbodiimide (Scheme 3, step 1 and 2). Furthermore, MiniKans can be mixed for all washing steps in between. Thereafter, MiniKans were divided, according to their peptoid sequence, between the different N-substituted glycine residues. Due to nucleophilic substitution the halide is displaced by the amine (Scheme 3, step 3). Afterwards further acylation with bromoacetic acid was done and the cycle was repeated three times, to obtain tetrameric peptoids (Scheme 3, step 4).

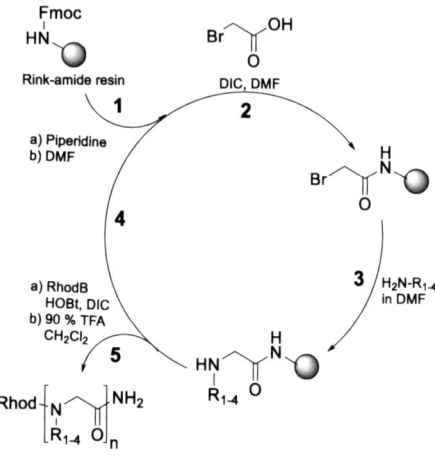

Scheme 3: Synthesis of peptoids by submonomer method. Amines used for synthesis are tert-butyl (4-aminobutyl)carbamate (R_1), prop-2-yn-1-amine (R_2), benzylamine (R_3) and 4-chlorobenzylamine (R_4).

To reveal a comprehensive library of peptoids, which can be analyzed for structure-function relationships, all positions in the tetrameric peptoid structure were completely permutated, thus the library was consisting of $4^4 = 256$ peptoids (sequences shown in table 17, appendix). Furthermore, all peptoids were coupled to rhodamine B, activated with 1-hydroxybenzotriazol (HOBt) and N,N'-diisopropylcarbodiimide (DIC), to facilitate the visualization of the peptoids in living cells and zebrafish embryos (Scheme 3, step 5a). Final cleavage from the resin (Scheme 3, step 5b) was done with 90% trifluoroacetic acid (TFA). This step was performed in a device, allowing simultaneous cleavage of up to 96 peptoids in MiniKans. The basic structure of a rhodamine B labeled tetrameric peptoid after final cleavage from the resin is shown in figure 6.

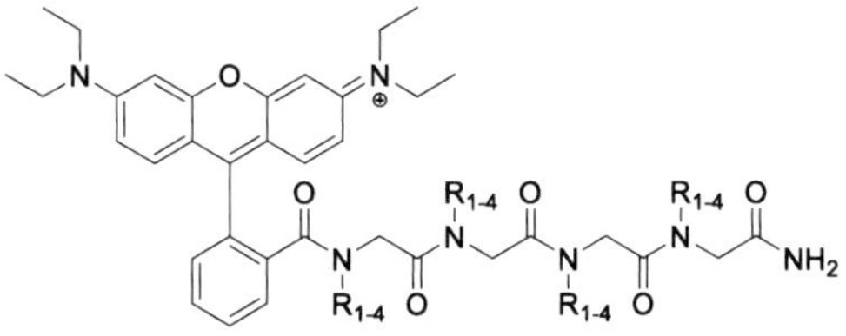

Figure 6: Basic structure of tetrameric peptoid labeled with rhodamine B.

In her thesis F. Rönicke used the unpurified peptoid library to increase the speed of the biological evaluation with the intention to only purify positive hits. However, it has been shown that side products, e.g. shorter incomplete products, led to false identification of intracellular localizations, making structure-function relationships for organelle specificity impossible [91]. Hence, the 256 peptoids were purified by HPLC and analyzed by MALDI-TOF MS, to revise cell permeability, intracellular localization and determine exact toxicity values. Finally, 234 out of 256 synthesized peptoids could successfully be identified in an average yield of 33%. To investigate optimal concentrations for live cell screening the cytotoxicity of representative peptoids in the library was determined in mammalian cells such as HeLa cervix carcinoma cells, using the MTT (3-(4,5-dimethylthiazol-2-yl)-2,5-diphenyltetrazolium bromide) assay. This assay is based on the reduction of the yellow tetrazolium salt to the purple formazan in living cells. After dissolving the formazan product, absorbance can be measured (595 nm) and metabolic activity of cells can be quantified. 24 peptoids containing all four side chains, without doublings, were tested for concentrations between 5 and 40 µM. In addition, four peptoids which only contain one side chain were investigated (figure 79, appendix). Most peptoids displayed no or only moderate toxicity in HeLa cells. For the majority of the peptoids LD_{50} values could not be determined as they were beyond the 40 µM ($LD_{50} >$ 40 µM). The only peptoid which displayed increased cytotoxicity was peptoid 86 (Npcb-Npcb-Npcb-Npcb-RhodB) with an LD_{50} value of about 30 µM. Subsequently, peptoids were screened for cellular uptake and localization in HeLa cells. 1.5 x 10^4 cells/well were seeded in 96 well µ-plates (IBIDI) and treated with 10 µM peptoid solution. After 24 h incubation the cells were washed with Dulbecco's phosphate-buffered saline (DPBS) and co-stained with 125 nM MitoTracker® Green and 2 µg/ml Hoechst 33342 to visualize mitochondria and cell nuclei. Cells were screened with a fully automated Olympus Scan^R Ix81 fluorescence microscope and afterwards image data was analyzed by a custom developed software script. In collaboration with Ralf Mikut and Markus Reischl (Institute for Automation and Applied Informatics, KIT), a MATLAB based script for automated quantification of cellular uptake and identification of mitochondrial localization of peptoids was developed [91]. First of all, artifacts, e.g. images out of focus, had to be excluded for further analysis. Cell nuclei and mitochondrial staining (green signal) were recognized and artifacts were identified by cluster analysis. Furthermore, remaining data was screened again manually to remove remaining images which were out of focus and not recognized by the automated approach. Recognition of cell nuclei for P82 and data analysis for the library is shown in figure 7, displaying artifacts in red (automated detection) and blue (manually detected) and remaining data points are represented in green.

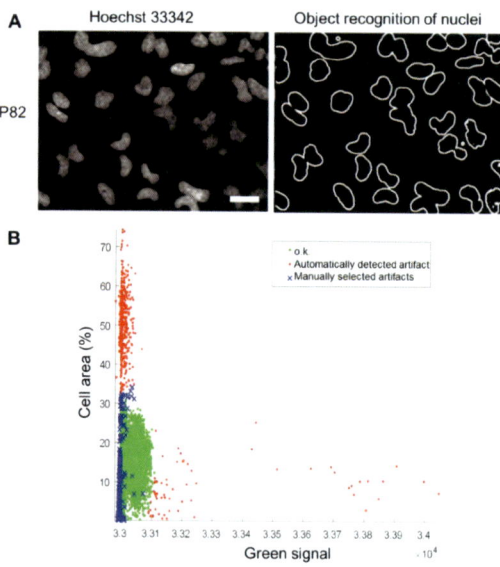

Figure 7: Artifact recognition *via* automated cluster analysis (red) and subsequent manual selection of artifacts (blue). Data points, for further analysis, are shown in green. Scale bar: 20 µm

Due to this evaluation, eight (P22, P41, P68, P70, P126, P132, P198 and P214) out of 234 screened peptoids (3.4%) had to be excluded for further investigation. A possible explanation, for exclusion of those peptoids, could be their toxicity. Hence, viability of cells, treated with peptoids (5 µM, 10 µM, 20 µM and 40 µM) for 72 h was determined using the MTT assay (figure 8).

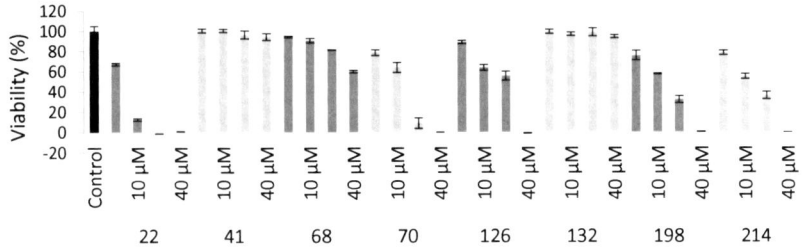

Figure 8: Cytotoxicity of peptoids leading to artifacts in the screening approach in HeLa cells: P22, P41, P68, P70, P126, P132, P198 and P214. Peptoids were tested in 5, 10, 20 and 40 µM concentration for 72 h and viability was determined using the MTT assay.

Increased toxicity was found for P22, P70, P126, P198 and P214 with LD_{50} values between 7.5 and 20 µM, leading to low cell density or dead cells in solution, hindering the automated microscope to focus. P41 and P132 displayed no toxicity in HeLa cells, even for 40 µM. Occurring artifacts for those peptoids could be by chance or due to low uptake of peptoids. Cells showing very low peptoid uptake were not detectable by the microscope, as the focus was adjusted by the peptoid signal.

The main focus was on peptoids accumulating in mitochondria, as mitochondria might play an important role in several diseases [112, 113]. Mitochondria penetrating peptoids (MPPos) are the counterparts to mitochondria penetrating peptides (MPPs), displaying high and effiecient uptake in mitochondria. For the identification of MPPos colocalization values of peptoids and MitoTracker® Green were quantified. While peptoids with high colocalization values accumulate in mitochondria, it is known that low colocalization values represent endosomal or lysosomal accumulation, as those peptoids display punctate patterns in the perinuclear region. Figure 9 shows the colocalization of two representative peptoids, for a mitochondrial and an endosomal peptoid, with the MitoTracker® Green signal. P82 (Npcb-Npcb-Nprg-Npcb-RhodB) is a representative MPPo, displaying a high overlap of red in green signal with a colocalization value of 87.6%, whereas low overlapping signals are shown for P123 (Npcb-Nphe-Nlys-Nlys-RhodB), an endosomal peptoid, with a colocalization value of 35.1%.

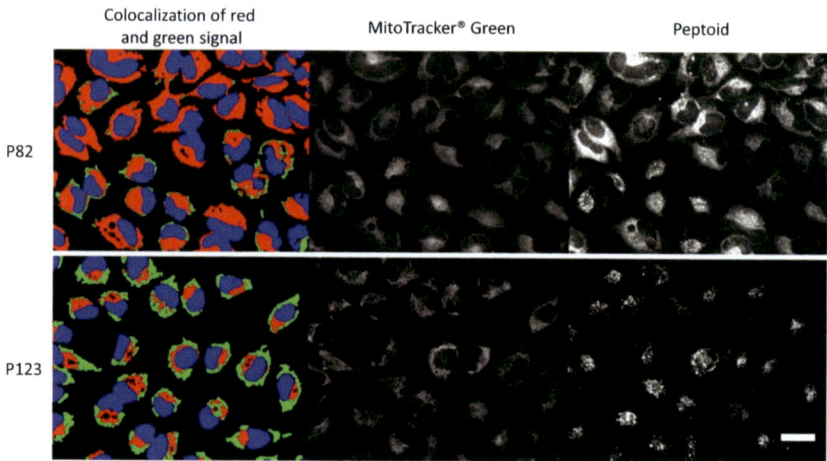

Figure 9: Automated colocalization recognition for P82 and P123. P82 shows a high colocalization ratio of red (peptoid) and green (MitoTracker® Green) signal whereas P123 displays a low colocalization value. Scale bar: 20 μm

As already shown by F. Rönicke, structure-function correlations for the different side chains, of the tetrameric peptoids, can be performed by investigation of colocalization values, as a function of the ratio of the respective side chain [91]. This analysis has been repeated with the purified library. Therefore, values were analyzed for 0% (absence of the particular side chain in the peptoid sequence), 25% (one residue of the particular side chain at any position in the peptoid sequence), 50% (two residues of the particular side chain at any position in the peptoid sequence), 75% (three residues of the particular side chain in the peptoid sequence) and 100% (all positions are containing the same side chain) for each moiety. Boxplots for Nprg (figure 10 A), Npcb (figure 10 B), Nlys (figure 10 C) and Nphe (figure 10 D) display significant effects for Npcb and Nlys. With an increasing ratio of Npcb, in the tetrameric peptoid, the colocalization of red and green signal was also rising. Hence, it can be assumed that the lipophilic side chain Npcb leads to mitochondrial accumulation. In contrast to that high ratios of Nlys led to decreasing colocalization values. Peptoids which were rich in the hydrophilic and cationic side chain Nlys displayed only low overlap with MitoTracker® Green signal. Therefore, they were probably accumulating in endosomes. For Nprg and Nphe a clear tendency could not be observed, assuming that their effect on the peptoid localization was rather low.

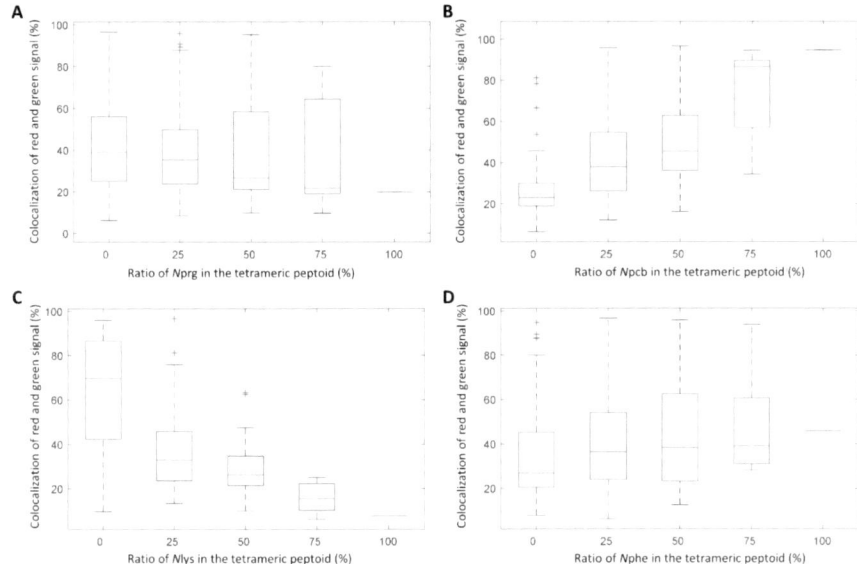

Figure 10: Box plot of the red- in green signal against the ratio of the respective side chain in the tetrameric peptoid (0, 25, 50, 75 and 100%). Box plots for Nprg (**A**), Npcb (**B**), Nlys (**C**) and Nphe (**D**) are shown.

The impact of side chains was not only analyzed on the intracellular localization but also on cellular uptake. Cellular uptake could be quantified by measuring the intensity of the rhodamine B fluorescence signal (red signal). High fluorescence intensities imply high and efficient cellular uptake of peptoids, whereas for peptoids with low cellular uptake the fluorescence signal is low. Box plots for red signals as a function of the ratio of the respective side chain in the peptoid are shown in figure 11. Significant results were found for Nprg, Npcb and Nlys. With an increasing ratio of Npcb in the peptoid the red signal was increasing. Hence, peptoids which are rich in Npcb displayed high cellular uptake while Nprg and Nlys led to decreased cellular uptake.

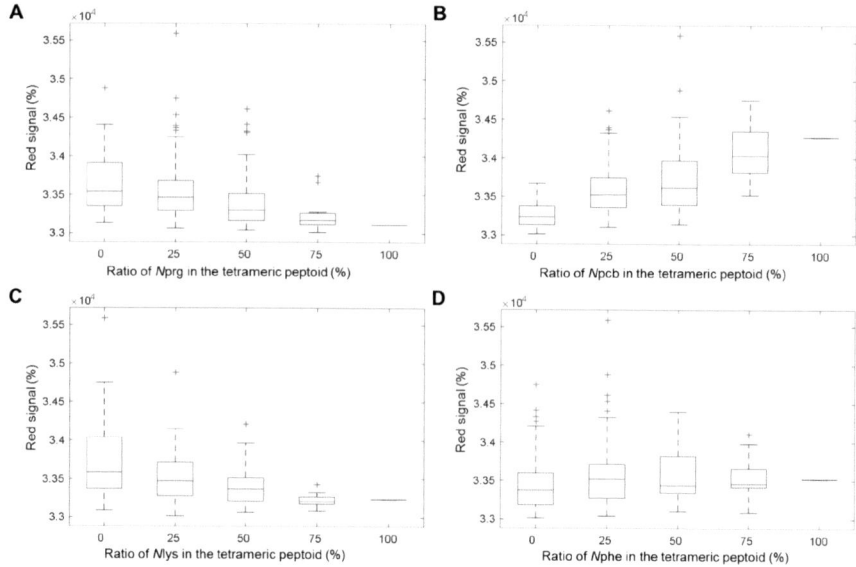

Figure 11: Box plots of the red signal against the ratio of the respective side chain in the tetrameric peptoid (0, 25, 50, 75 and 100%). Box plots for *N*prg (**A**), *N*pcb (**B**), *N*lys (**C**) and *N*phe (**D**) are shown.

Additionally, peptoids were subdivided in three subgroups according to their colocalization values in order to analyze structure-function relationship for each peptoid in the library. 226 peptoids were successful screened for cellular uptake in HeLa cells and displayed accumulation in endosomes (colocalization value <50%), endosomes and mitochondria (colocalization value 50 – 59%) or mitochondria (colocalization value >59%). The majority of peptoids was found in the endosomal subgroup, containing 163 peptoids. The most occurring side chain in this group was *N*lys with 33% and least appearing is *N*pcb with 20%. Especially, the 10 peptoids with the lowest colocalization values in the library were rich in *N*lys (65%) and contained only 2% *N*pcb (figure 12, row 1). The subgroup for peptoids accumulating in endosomes and mitochondria represents the smallest subgroup with 16 peptoids. Peptoids in this subgroup were accumulating in both organelles or exclusively in mitochondria or endosomes and therefore need to be screened manual again and classification has to be done by visual identification of peptoid signal. They contained a mixture of hydrophobic and hydrophilic side chains and the most occurring one was *N*phe, which was already found to have no significant effects on intracellular peptoid localization. The subgroup for MPPos contained 47 peptoids and 94% of those contained at least one-time Npcb (overall average 40%). In contrast, only nine peptoids contained Nlys (overall average 6%). Nprg and Nphe were present with 25% and 29%. These findings confirm, that lipophilic peptoids, which are rich in Npcb, display mitochondrial uptake.

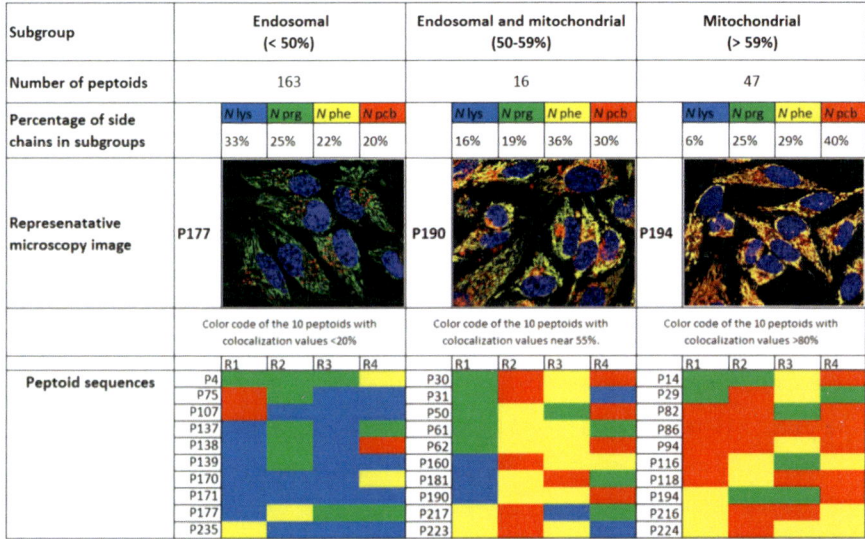

Figure 12: Overview of the subgroup classification of the peptoid library after automated screening and evaluation in endosomal, endosomal and mitochondrial and mitochondrial peptoids and their corresponding side chain content. Merged colocalization images of subgroup representatives P31, P190 and P194. Below, ten selected sequences for the respective subgroup are shown in their color code.

Furthermore, image data were checked manually to verify successful classification of the automated qualification and identification of MMPos. In the subgroup for MPPos (colocalization value >59%) three peptoids, P101 (Npcb-Nlys-Npcb-Nprg-RhodB), P103 (Npcb-Nlys-Npcb-Nlys-RhodB) and P166 (Nlys-Nlys-Npcb-Npcb-RhodB) which displayed predominate endosomal uptake were found. A possible reason for false automated classification of P101, P103 and P166 could be increased cellular uptake due to high ratio of Npcb. Strong endosomal uptake, not only in perinuclear regions but all over the cell, can lead to overlap of endosomal peptoid signal with MitoTracker® Green signal and therefore increased colocalization values. In the subgroup for endosomal peptoids six MPPos (P24, P80, P125, P208, P244 and P256) were found. The structure of these MPPos could provide a supposable reason for their false automated classification. These peptoids were rich in Nphe (63%), a side chain which does not enhance cellular uptake compared to Npcb (figure 10). Hence, mitochondrial uptake was comparable low and due to this peptoids were false recognized as endosomal peptoids.

Peptoid	Sequence	Classification automated	Classification manual
P24		Endosomal	Mitochondrial
P80		Endosomal	Mitochondrial
P101		Mitochondrial	Endosomal
P103		Mitochondrial	Endosomal
P125		Endosomal	Mitochondrial
P166		Mitochondrial	Endosomal
P208		Endosomal	Mitochondrial
P244		Endosomal	Mitochondrial
P256		Endosomal	Mitochondrial

Figure 13: Peptoids which were classified to the false subgroup, as found after visual examination, in the automated process. Peptoid numbers, sequences and automated and manual classifications are shown.

The subgroup for endosomal and mitochondrial peptoids contained a mixture of MPPos and endosomal peptoids as well as peptoids accumulating in both organelles. Manual classification revealed six MPPos (P30, P50, P61, P62, P210, 245), four endosomal peptoids (P31, P104, P160 and P223) and six peptoids with no clear localization (P71, P181, P190, P217, P220 and P247). In total, automated image analysis and classification was successful with only nine peptoids, shown in figure 13, assigned to the false subgroup, revealing a success rate of 96%.

3.1.1.1. Octanol-water partition coefficients

In order to further analyze the structure-function relationship for the prediction of mitochondria localizing peptoids, octanol-water partition coefficients (K_{ow}) were measured. Therefore, the ratio of a compound concentration in octanol is compared to its concentration in water. The observed coefficients are very important in pharmacology to understand accumulation and distribution of a drug in cells or whole organisms. Furthermore, it can pre-estimate clearance, dosing and toxicity of drugs. Usually the logarithm of the ratio is used:

$$LogP = log \frac{Concentration\ (octanol)}{Concentration\ (water)}$$

While compounds which are highly water-soluble display LogP values below zero, lipophilic compounds, accumulating in the octanol phase, exhibit values above zero. According to Lipinskis's rule of five, which describe optimal properties for drugs, LogP values should not be greater than 5 and are optimal between -0.4 and 5.6 [114, 115]. Compounds with values greater than 5 are too lipophilic to be soluble in water and therefore display low bioavailability. On the other hand, highly water-soluble compounds, with values above zero, are not able to cross hydrophobic cell membranes. It has been shown that optimal LogP values for CNS specific drugs are around 2 [116]. Heroin, for example, which is able to cross the blood-brain-barrier has a LogP value of 1.58 (experimental) [117]. In comparison to that morphine which is three times less potent also displays lower hydrophobicity with a LogP value of 0.89 [117]. There is not only a correlation between LogP values and organ specificity. The LogP value can also used to predict intracellular localizations. For peptides it has been shown, that the LogP value is suitable to distinguish between mitochondrial and endosomal localization [118]. To investigate LogP values of CPPos, five representative peptoids were chosen for each subgroup. P75, P107, P137, P170 and P177 were clearly located in endosomes, P50, P160, P190, P217 and P223 were accumulating in endosomes and mitochondria and P82, P94, P116, P118 and P216 are representative MPPos. LogP values were determined by dissolving 80 µM peptoid in octanol-water solution (1:1). Absorbance of rhodamine B labeled peptoids in the octanol- and water phase was measured at 560 nm and the concentrations were calculated with Lambert-Beer law:

$$c = \frac{A}{\varepsilon * d}$$

Accumulation of CPPos in octanol (upper phase) and water (lower phase) is shown in figure 14, representing clearly different solubility's of peptoids in different subgroups.

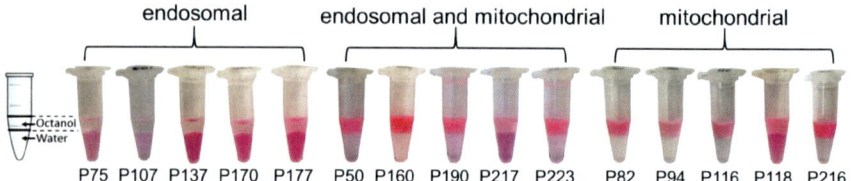

Figure 14: Determination of octanol-water partition coefficients of five endosomal, five endosomal and mitochondrial and five mitochondrial peptoids. Peptoids are solved in water and octanol (1:1) in eppendorf tubes (octanol: upper layer, water: lower layer) and enrichment of peptoids can be visual determined by magenta-colored solution.

Endosomal peptoids were mainly accumulating in water phase and mitochondrial peptoids were, due to higher lipophilicity, in octanol phase. Peptoids which could not be clearly classified to one or the other organelle, by automated image analysis, were accumulating in both phases. Experimental determined LogP values are presented in table 3, as well as theoretical LogP values predicted by cheminformatics webpage *Molinspiration* [119].

Table 3: LogP values of five endosomal, five endosomal and mitochondrial and five mitochondrial peptoids. LogP values were determined experimental by measuring the octanol-water coefficient and theoretical values were calculated using *Molinspiration*.

Peptoid	Subgroup	LogP experimental	LogP theoretical
P75	endosomes	-1.30	-2.32
P107	endosomes	-2.51	-2.04
P137	endosomes	-1.72	-3.48
P170	endosomes	-1.13	-1.64
P177	endosomes	-0.78	-0.92
P50	endosomes and mitochondria	1.96	1.95
P160	endosomes and mitochondria	1.05	0.24
P190	endosomes and mitochondria	-0.24	0.24
P217	endosomes and mitochondria	-0.43	-0.24
P223	endosomes and mitochondria	1.06	0.24
P82	mitochondria	2.71	4.55
P94	mitochondria	2.22	5.79

P116	mitochondria	2.89	3.19
P118	mitochondria	2.13	7.22
P216	mitochondria	2.82	5.11

LogP values confirm visual findings, that endosomal peptoids are water soluble with LogP values < -1. Theoretical predicted values are slightly different, however, they display the same tendency. MPPos were more lipophilic with LogP values between 2.13 and 2.89 (experimental) and even higher values in theoretical prediction (2.6 – 7.22). CPPos which were found in both compartments displayed LogP values in-between. These findings suit very well to determination of LogP values of CPPs and MPPs [118].

Another suitable prediction of behavior of CPPs and CPPos in octanol and water are molecular dynamic simulations [120]. It can be investigated how different side chains in the peptoid behave in water and octanol and the three-dimensional structure of the molecule in the respective solvent can be analyzed. In cooperation with Daniel Holub (Institute of Physical Chemistry, KIT) a representative endosomal peptoid (P170: *N*lys-*N*lys-*N*lys-*N*pcb-RhodB) and mitochondrial peptoid (P216: *N*phe-*N*pcb-*N*pcb-*N*phe-RhodB) were simulated in an unbiased molecular dynamic simulation for 300 ns in octanol and water, using GROMACS software tools. Snapshots of both peptoids were taken after 300 ns, showing the localization of the peptoid within the octanol (blue) and water (red) box (figure 15).

P170   P216

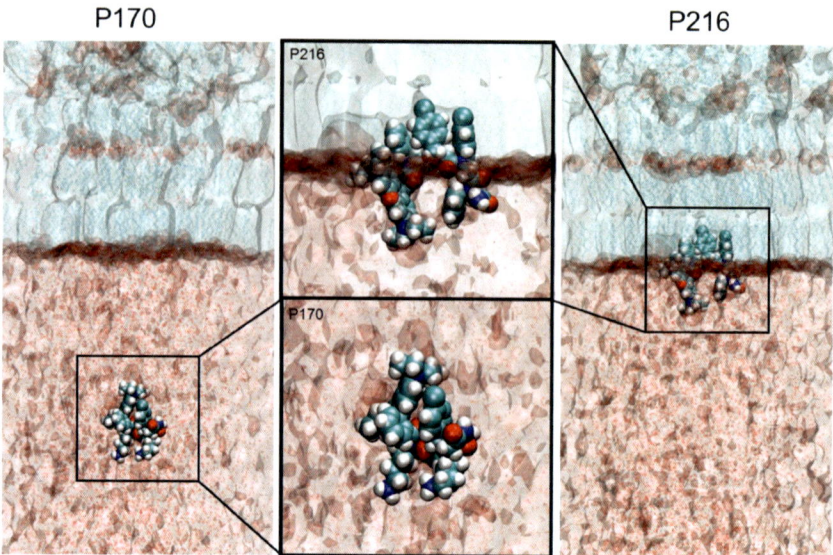

Figure 15: Molecular dynamic simulation of endosomal peptoid P170 (*N*lys-*N*lys-*N*lys-*N*pcb-RhodB) and mitochondrial peptoid P216 (*N*phe-*N*pcb-*N*pcb-*N*phe-RhodB). Octanol phase is represented in blue and water is represented in red, carbon atoms are shown in light blue, oxygen in red, nitrogen in dark blue and hydrogen in white. Snapshots were taken after 300 ns and locations, within the solvents, of peptoids are shown (P216: left and upper middle part, P170: right and lower middle part).

The hydrophobic peptoid P216 was located at the border between the octanol and water phase, extending its side chains (*N*phe and *N*pcb) to the octanol and the more hydrophilic and charged rhodamine B, the oxygen atoms of the backbone, as well as the amide bonding of the C-terminus were presented to the water phase. The hydrophilic peptoid was found in the water phase with a bended backbone, extending its positive charged side chains to the organic solvent.

3.1.1.2. Cytotoxicity of MPPos

In total, 56 MPPos were found after automated and subsequently visual screening of peptoid library 1 (44 MPPos detected automatically, 6 MPPos in endosomal and mitochondrial subgroup, and 6 MPPos mismatched to endosomal subgroup). MPPos are especially interesting for drug delivery as mitochondria are involved in many essential processes, including cellular metabolism, apoptosis and cell signaling [121-124]. Mitochondria might play an important role in diseases, such as cancer or neurological disorders [112, 113]. For the usage of MPPos as molecular transporters for mitochondria active drugs, in order to transport different cargos into cells, it is important that they display a low cytotoxicity. As mitochondria regulate cellular metabolism and produce ATP *via* cellular respiration, it is important that MPPos do not disturb cellular function by high accumulation in mitochondria. Therefore, MTT assays were performed for all 56 MPPos, using concentrations in low micromolar range (5, 10, 20 and 40 µM) for 72 h in HeLa cells, as shown in figure 16.

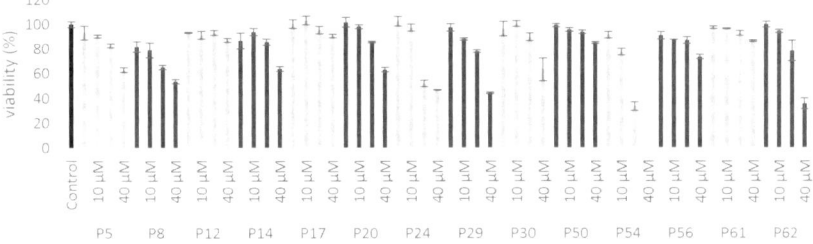

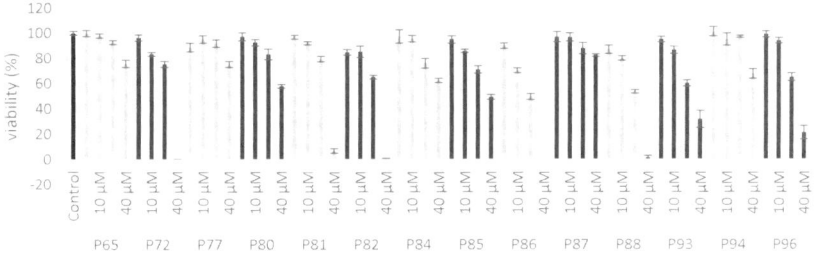

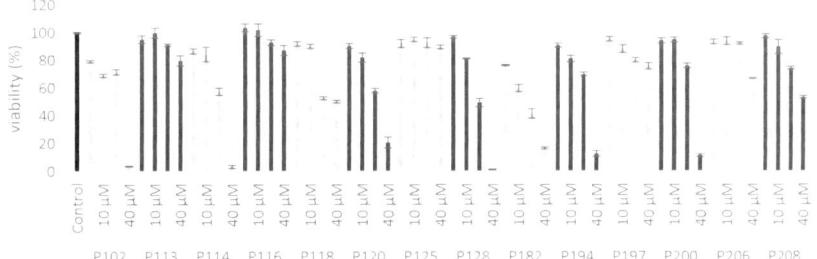

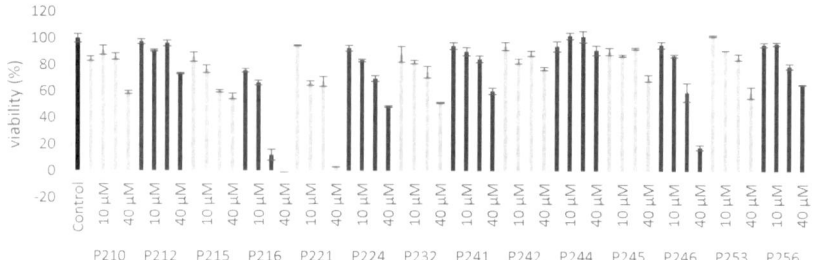

Figure 16: Cytotoxicity of MPPos in HeLa cells. Peptoids were tested in 5, 10, 20 and 40 µM concentrations for 72 h and viability was determined using the MTT assay.

Evaluation of MTT assays revealed 33 MPPos with no or modearte cytotoxicity and 23 MPPos with increased cytotoxicty in HeLa cells and LD_{50} values between 13 and 40 µM. MPPos which displayed low cytotoxicity are suitable molecular transporters and the most occuring side chain was Nphe, with 40%. Nprg was the second frequently side chain with 33% and Npcb was only occuring with 24%. Only four peptoids contained the hydrophilic and cationic side chain Nlys (3%). Most of the toxic MPPos were very lipophilic with 51% Npcb and each peptoid contained this side chain at least once. Nphe and Nprg were present with 29% and 17% and they contained only 2% Nlys. Consequently, high ratios of Npcb in tertrameric peptoids were leading to enhanced mitochondrial uptake. However, peptoids with more than 50% Npbc displayed increased toxicity in HeLa cells and might not be suitable as transporter molecules.

3.1.2. Peptoid library 2

In collaboration with Bettina Fleck (Institute of Toxicology and Genetics, KIT) a second tetrameric peptoid library was synthesized. To increase the amount of MPPos, three aromatic polar and non-polar side chains, shown in figure 17, were chosen: benzylamine (R₃, Nphe, yellow), fluorobenzylamine (R₆, Npfb, black), and 4-hydroxybenzylamine. (R₅, Npob, magenta). Npob represents the natural, polar amino acid tyrosine and Npbf was chosen to replace Npcb, which induced increased cytotoxicity as shown in chapter 3.1.1.3.. As Npbf (LogP ~ 1.28) is a non-polar side chain but less lipophilic than Npbc (LogP ~ 1.8), it might enhance mitochondrial uptake without increasing the toxicity of peptoids to cells or whole organism.

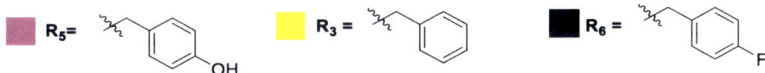

Figure 17: Amines used for synthesis of peptoid library 2: 4-hydroxybenzylamine (pink, R₅), benzylamine (yellow, R₃) and 4-fluorobenzylamine (black, R₆).

To reveal a fully permutated library 3^4 = 81 peptoids were synthesized by submonomer method in IRORI MiniKans as described for library 1. Subsequently, all CPPos were labeled with rhodamine B and after cleavage with 95% TFA identified by mass spectrometry. 74 peptoids were successfully identified and peptoids sequences of library 2 as well as their respective molecular weights are shown in table 4.

Table 4: Overview of peptoids in library 2. Peptoid numbers, sequences and respective molecular weights (g/mol) are shown. Success of synthesis was verified by MALDI-TOF mass spectrometry and identified peptoids are marked with "x".

Peptoid	Sequence	Molecular weight (g/mol)	Mass found (x)
P257	Nphe-Nphe-Nphe-Nphe-RhodB	1030.61	X
P258	Nphe-Nphe-Nphe-Npfb-RhodB	1049.28	X
P259	Nphe-Nphe-Nphe-Npob-RhodB	1047.29	X
P260	Nphe-Nphe-Npbf-Nphe-RhodB	1049.28	X
P261	Nphe-Nphe-Npbf-Npbf-RhodB	1067.27	X
P262	Nphe-Nphe-Npbf-Npob-RhodB	1065.28	X
P263	Nphe-Nphe-Npob-Nphe-RhodB	1047.29	X
P264	Nphe-Nphe-Npob-Npbf-RhodB	1065.28	X

P265	*N*phe-*N*phe-*N*pob-*N*pob-RhodB	1063.29	X
P266	*N*phe-*N*pbf-*N*phe-*N*phe-RhodB	1049.28	X
P267	*N*phe-*N*pbf-*N*phe-*N*pbf-RhodB	1067.27	X
P268	*N*phe-*N*pbf-*N*phe-*N*pob-RhodB	1065.28	X
P269	*N*phe-*N*pbf-*N*pbf-*N*phe-RhodB	1067.27	X
P270	*N*phe-*N*pbf-*N*pbf-*N*pbf-RhodB	1085.26	X
P271	*N*phe-*N*pbf-*N*pbf-*N*pob-RhodB	1083.27	X
P272	*N*phe-*N*pbf-*N*pob-*N*phe-RhodB	1065.28	X
P273	*N*phe-*N*pbf-*N*pob-*N*pbf-RhodB	1083.27	X
P274	*N*phe-*N*pbf-*N*pob-*N*pob-RhodB	1081.28	X
P275	*N*phe-*N*pob-*N*phe-*N*phe-RhodB	1047.29	X
P276	*N*phe-*N*pob-*N*phe-*N*pbf-RhodB	1065.28	X
P277	*N*phe-*N*pob-*N*phe-*N*pob-RhodB	1063.29	X
P278	*N*phe-*N*pob-*N*pbf-*N*phe-RhodB	1065.28	X
P279	*N*phe-*N*pob-*N*pbf-*N*pbf-RhodB	1083.27	X
P280	*N*phe-*N*pob-*N*pbf-*N*pob-RhodB	1081.28	X
P281	*N*phe-*N*pob-*N*pob-*N*phe-RhodB	1063.29	X
P282	*N*phe-*N*pob-*N*pob-*N*pbf-RhodB	1081.28	X
P283	*N*phe-*N*pob-*N*pob-*N*pob-RhodB	1079.29	-
P284	*N*pbf-*N*phe-*N*phe-*N*phe-RhodB	1049.28	X
P285	*N*pbf-*N*phe-*N*phe-*N*pfb-RhodB	1067.27	X
P286	*N*pbf-*N*phe-*N*phe-*N*pob-RhodB	1065.28	X
P287	*N*pbf-*N*phe-*N*pbf-*N*phe-RhodB	1067.27	X
P288	*N*pbf-*N*phe-*N*pbf-*N*pbf-RhodB	1085.26	X
P289	*N*pbf-*N*phe-*N*pbf-*N*pob-RhodB	1083.27	X
P290	*N*pbf-*N*phe-*N*pob-*N*phe-RhodB	1065.28	X
P291	*N*pbf-*N*phe-*N*pob-*N*pbf-RhodB	1083.27	X
P292	*N*pbf-*N*phe-*N*pob-*N*pob-RhodB	1081.28	X
P293	*N*pbf-*N*pbf-*N*phe-*N*phe-RhodB	1067.27	X
P294	*N*pbf-*N*pbf-*N*phe-*N*pbf-RhodB	1085.26	X
P295	*N*pbf-*N*pbf-*N*phe-*N*pob-RhodB	1083.27	X
P296	*N*pbf-*N*pbf-*N*pbf-*N*phe-RhodB	1085.26	X

P297	*N*pbf-*N*pbf-*N*pbf-*N*pbf-RhodB	1103.25	-
P298	*N*pbf-*N*pbf-*N*pbf-*N*pob-RhodB	1101.26	X
P299	*N*pbf-*N*pbf-*N*pob-*N*phe-RhodB	1083.27	X
P300	*N*pbf-*N*pbf-*N*pob-*N*pbf-RhodB	1101.26	X
P301	*N*pbf-*N*pbf-*N*pob-*N*pob-RhodB	1099.27	X
P302	*N*pbf-*N*pob-*N*phe-*N*phe-RhodB	1065.28	X
P303	*N*pbf-*N*pob-*N*phe-*N*pbf-RhodB	1083.27	X
P304	*N*pbf-*N*pob-*N*phe-*N*pob-RhodB	1081.28	X
P305	*N*pbf-*N*pob-*N*pbf-*N*phe-RhodB	1083.27	X
P306	*N*pbf-*N*pob-*N*pbf-*N*pbf-RhodB	1101.26	X
P307	*N*pbf-*N*pob-*N*pbf-*N*pob-RhodB	1099.27	X
P308	*N*pbf-*N*pob-*N*pob-*N*phe-RhodB	1081.28	X
P309	*N*pbf-*N*pob-*N*pob-*N*pbf-RhodB	1099.27	X
P310	*N*pbf-*N*pob-*N*pob-*N*pob-RhodB	1097.28	X
P311	*N*pob-*N*phe-*N*phe-*N*phe-RhodB	1047.29	X
P312	*N*pob-*N*phe-*N*phe-*N*pfb-RhodB	1065.28	X
P313	*N*pob-*N*phe-*N*phe-*N*pob-RhodB	1063.29	x
P314	*N*pob-*N*phe-*N*pbf-*N*phe-RhodB	1065.28	X
P315	*N*pob-*N*phe-*N*pbf-*N*pbf-RhodB	1083.27	X
P316	*N*pob-*N*phe-*N*pbf-*N*pob-RhodB	1081.28	X
P317	*N*pob-*N*phe-*N*pob-*N*phe-RhodB	1063.29	X
P318	*N*pob-*N*phe-*N*pob-*N*pbf-RhodB	1081.28	X
P319	*N*pob-*N*phe-*N*pob-*N*pob-RhodB	1079.29	X
P320	*N*pob-*N*pbf-*N*phe-*N*phe-RhodB	1065.28	X
P321	*N*pob-*N*pbf-*N*phe-*N*pbf-RhodB	1083.27	-
P322	*N*pob-*N*pbf-*N*phe-*N*pob-RhodB	1081.28	-
P323	*N*pob-*N*pbf-*N*pbf-*N*phe-RhodB	1083.27	X
P324	*N*pob-*N*pbf-*N*pbf-*N*pbf-RhodB	1101.26	-
P325	*N*pob-*N*pbf-*N*pbf-*N*pob-RhodB	1099.27	X
P326	*N*pob-*N*pbf-*N*pob-*N*phe-RhodB	1081.28	X
P327	*N*pob-*N*pbf-*N*pob-*N*pbf-RhodB	1099.27	X
P328	*N*pob-*N*pbf-*N*pob-*N*pob-RhodB	1097.28	X

P329	Npob-Npob-Nphe-Nphe-RhodB	1063.29	X
P330	Npob-Npob-Nphe-Npbf-RhodB	1081.28	X
P331	Npob-Npob-Nphe-Npob-RhodB	1079.29	X
P332	Npob-Npob-Npbf-Nphe-RhodB	1081.28	X
P333	Npob-Npob-Npbf-Npbf-RhodB	1099.27	-
P334	Npob-Npob-Npbf-Npob-RhodB	1097.28	X
P335	Npob-Npob-Npob-Nphe-RhodB	1079.29	X
P336	Npob-Npob-Npob-Npbf-RhodB	1097.27	X
P337	Npob-Npob-Npob-Npob-RhodB	1095.29	-

Afterwards, the peptoids of library 2 were screened for cellular uptake and intracellular localization in HeLa cells as shown for library 1 (chapter 3.1.). After incubation of 10 µM peptoid solution for 24 h in 1.5 x 10⁴ HeLa cells/well, cells were washed with DPBS and co-stained with 125 nM MitoTracker® Green and 2 µg/ml Hoechst 33342. Likewise, cells were screened automatically by using an Olympus Scan^R Ix81 fluorescence microscope. All peptoids displayed strong cellular uptake and in comparison to library 1 the amount of occurring artifacts in the screening process was low. Whereas in library 1 eight peptoids had to be excluded, mainly due to increased toxicity, all peptoids in library 2 revealed microscopy images in good quality, sufficient for further analysis. As expected, library 2 contained mainly MPPos showing strong uptake and high colocalization with MitoTracker® Green. Consistently to library 1, localization of peptoids in library 2 can also be subdivided in three groups: endosomal, endosomal and mitochondrial and mitochondrial uptake. However, while the mitochondrial subgroup of library 1 was rather small, including only 26% of all analyzed peptoids, 66% of peptoids in library 2 were found in this subgroup. The most occurring side chain in this group was Nphe with 41%, which was present in most peptoids at least once (91%). Npbf was present with 37%, also occurring in 89% of MPPos. Most unfrequently was Npob with 22%. It was occurring in 73% MPPos, however, only 9% contained it more than once.

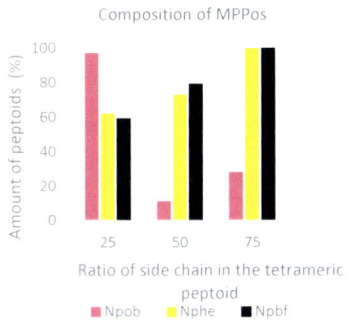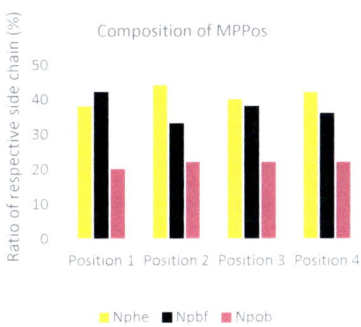

Figure 18: Analysis of the composition of MPPos in library 2. Left: Amount of peptoids (%) containing the respective side chain (Npob, Nphe, Npbf) once (25%), twice (50%) or thrice (75%) in MPPos in library 2. Right: Content of the respective side chain at position 1-4 in the tetrameric peptoid.

Furthermore, the exact composition of MPPos was determined, as shown in figure 18, and accumulation of peptoids containing the respective side chain once (ratio = 25%), twice (ratio = 50%) or thrice (ratio = 75 %) was analyzed. Amounts of peptoids in each subgroup (25%, 50% and 75%) were normalized to 100%. Most peptoids containing Npob only once were accumulating in mitochondria (97%). However, the amount of peptoids containing Npob twice or three times was very low. For Npbf and Nphe, it was the contrary. About 40% of peptoids containing Npbf/Nphe only once were absent in the mitochondrial subgroup. However, most peptoids containing Npbf/Nphe twice (~75%) and thrice (100%) were present in this subgroup. Figure 18 right shows the ratio of side chain in relation to the position in the peptoid. The impact of the exact position of the side chain in the peptoid seemed not to be essential for mitochondrial localization. The proportion of each side chain was consistent within the peptoid. Only on position one, ratio of Npbf was lower compared to Nphe. A possible explanation for the low impacts of the side chain position could be the short size of the peptoids as well as their high flexibility.

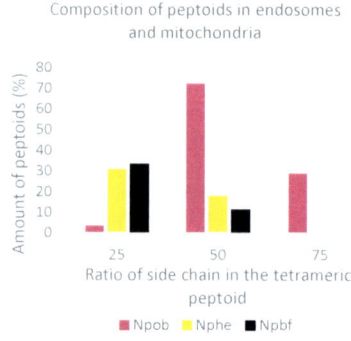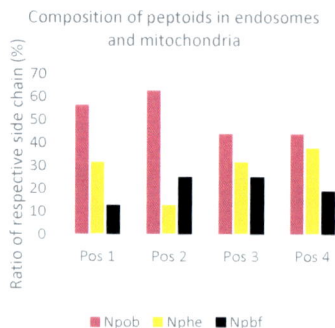

Figure 19: Analysis of the composition of peptoids of library 2 accumulating in endosomes and mitochondria. Left: Amount of peptoids (%) containing the respective side chain (*Npob*, *Nphe*, *Npbf*) once (25%), twice (50%) or thrice (75%) in endosomal and mitochondrial peptoids in library 2. Right: Content of the respective side chain at Position 1-4 in the tetrameric peptoid.

The subgroup of endosomal/mitochondrial peptoids contained 16 peptoids. The most frequent side chain was *Npob* with 52%. *Npob* was present in each peptoid and 94% of the peptoid in this subgroup contained *Npob* at least twice. *Nphe* was present with 28% and lowest occurring side chain was *Npbf* with 20%. Figure 19 shows that most peptoids containing *Npob* twice (72%) are present in this group. Furthermore, 29% of the peptoids containing *Npob* three times were in this group and only 3% of the peptoids contained *Npob* only once. A low amount of peptoids containing *Nphe* once (30%) or twice (18%) were present in this group, as well as peptoids containing *Npbf* once (33%). Only one peptoid containing *Npbf* twice was found in this group. Peptoids with three *Nphe* or *Npbf* residues could not be found in this group.

The group of endosomal peptoids was very small, as only seven peptoids (10%) showed uptake by endocytosis. Those peptoids were rich in *Npob* (61%), which was present in each peptoid at least two times, and non-polar side chains *Npbf* and *Nphe* were less frequent with 24% and 14%.

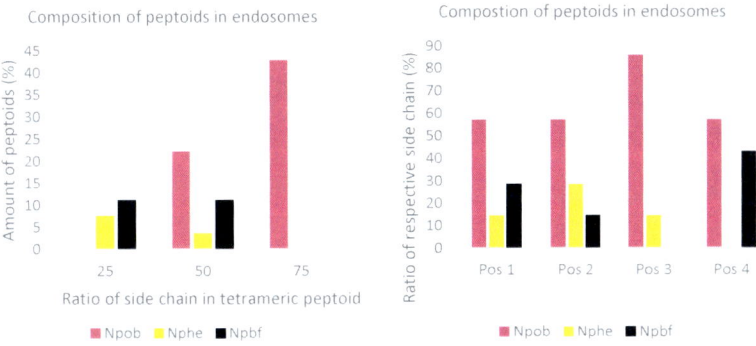

Figure 20: Analysis of the composition of peptoids of library 2 accumulating in endosomes. Left: Amount of peptoids (%) containing the respective side chain (Npob, Nphe, Npbf) once (25%), twice (50%) or thrice (75%) in endosomal peptoids of library 2. Right: Content of the respective side chain at Position 1-4 in the tetrameric peptoid.

Most peptoids which contain Npob thrice were present in this group and strongly lipophilic peptoids with more than one Nphe or Npbf were rare (figure 20). The side chains were not as equally distributed as shown for the previous groups. Especially in position 3 the amount of Npob was very high. While Npbf was frequently occurring in position four, Nphe could not be found on this position (figure 20). However, the endosomal group was very small and therefore detailed structure-function relationship analysis were difficult.

Confocal microscopy images of two representative peptoids with different localizations, P294 (Npbf-Npbf-Nphe-Npbf-RhodB) and P319 (Npob-Nphe-Npob-Npob-RhodB), are shown in figure 21. P294 is rich in Npbf, thus high colocalization with MitoTracker® Green was found. In contrast, P319 is less lipophilic and endocytotic uptake, with low colocalization to the MitoTracker® Green, could be observed.

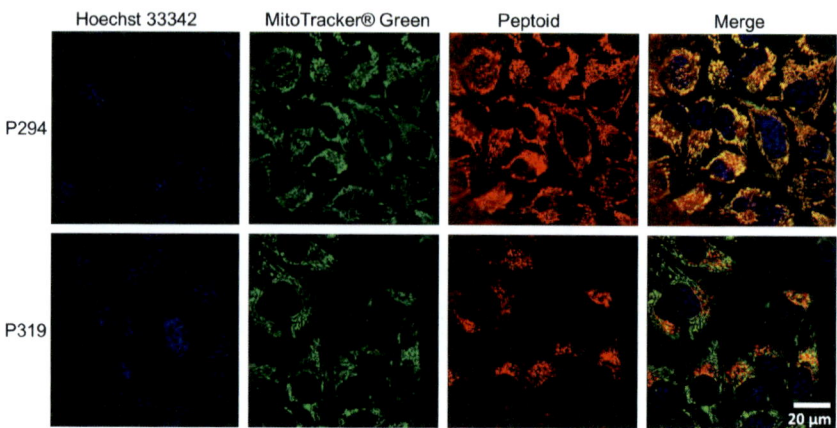

Figure 21: Cellular uptake of peptoids (P294 (*Npbf-Npbf-Nphe-Npbf*-RhodB), P319 (*Npob-Nphe-Npob-Npob*-RhodB) in HeLa cells. 1.5×10^4 cells were treated with 10 µM P294 and P319 for 24 h at 37 °C. For co-staining of mitochondria and nuclei cells were treated with 125 nM MitoTracker® Green and Hoechst 33342 (2 µg/ml). Intracellular accumulation of the peptoids was detected with fluorescence confocal microscopy (Leica TCS-SPE, Objective: ACS APO 63x/1.30 OIL). 1. Hoechst 33342, Ex.: 405 nm, Em.: 417-468 nm, 2. MitoTracker® Green Ex.: 488 nm, Em.: 499-552 nm 3. Peptoid, Ex.: 561 nm, Em.: 593-696 nm 4. Merge, Scale bar: 20 µm

3.1.2.1. Cytotoxicity of MPPos in library 2

Ten exemplary MPPos of library 2 were tested for cytotoxic effects, as low toxicity of MPPos is critical for drug delivery (chapter 3.1.1.3.). P284, P286, P287, P296, P298, P310, P311, P316, P335 were chosen to cover all available compositions of ratios for the respective side chain in the tetrameric peptoid. Thus, peptoids containing *N*phe, *N*pob and *N*pbf once, twice or thrice except for a peptoid containing *N*pob thrice, as those peptoids are not located in mitochondria (chapter 3.1.2.). 0.75×10^4 HeLa cells/well were treated with 5, 10, 20 and 40 µM peptoid for 72 h and subsequently viability was determined using the MTT assay, shown in figure 22. LD_{50} values could be determined for P284 (~ 20 µM), P287 (~ 30 µM) and 296 (~ 10 µM). In contrast to the other seven tested peptoids, P284 (*Npbf-Nphe-Nphe-Nphe*-RhodB), P287 (*Npbf-Nphe-Npbf-Nphe*-RhodB) and P296 (*Npbf-Npbf-Npbf-Nphe*-RhodB) consist only of *N*phe and *N*pbf. Enhanced toxicity of P296 was probably due to its high content of *N*pbf, leading to increased lipophilicity and therefore increased mitochondria accumulation. All other peptoids, containing *N*pob at least once, displayed no toxic effects or only moderate toxicity even for

40 µM. Even for P298, containing *N*pbf thrice, and P311, containing *N*phe thrice, the cytotoxicity was low.

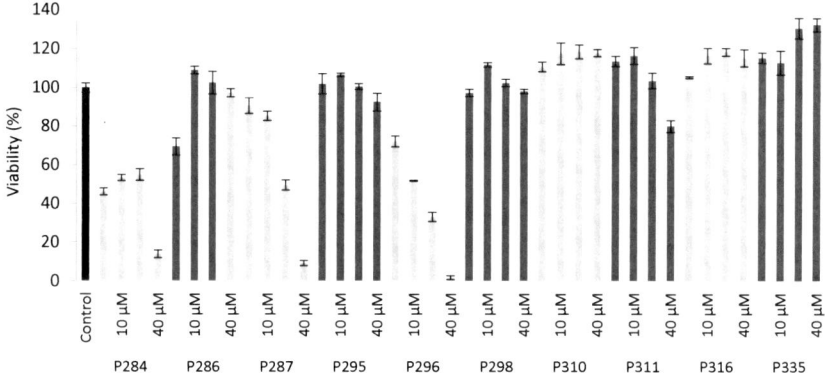

Figure 22 : Cytotoxicity of exemplary MPPos in library 2 counterpart in HeLa cells. Peptoids (P284, P286, P287, P295, P296, P298, P310, P311, P316 and P335) were tested in 5, 10, 20 and 40 µM concentration for 72 h and viability was determined using the MTT assay.

These findings confirm that increased lipophilicity is leading to increased cytotoxicity, as found for library 1. The side chain *N*pbf was used as lipophilic component, which might be less toxic compared to *N*pcb. However, a significant decrease of toxcity for peptoids rich in *N*pbf compared to *N*pcb could not be determined. Furthermore, these results highlight *N*pob as suitable side chain for MPPos as drug delivery systems. Peptoids containing *N*pob once and a mixture of *N*phe and *N*pbf displayed high colocalization with the MitoTracker® Green signal and also low toxcicity.

3.1.3.   Peptoid library 3

So far, only tetrameric CPPo libraries were analyzed for cell penetration and intracellular localization. For large molecules, such as proteins, tetrameric peptoids might be to small in comparison to their cargo, and not be sufficient for delivery of the cargo into cells. Thus, a small library of decameric and dodecameric CPPos was synthezised. As this library contains only eight CPPos and split-mix synthesis was not necessary, all peptoids were synthesized by submonomer method, using a fully automated peptoid synthesizer (Molecular Foundry, Lawrence Berkeley National Laboratory). Fully automated synthesis also gives the possibility to increase the amount of the synthesized peptoids. For library 3, a mixture of hydrophilic and hydrophopbic peptoids was used with *N*lys in each peptoid, to ensure water solubility and cell penetration (figure 23). Large CPPos, containing only lipophilic residues, would display high octanol-water partition coefficients and according to Lipinskis's rule of five, substances with LogP values greater than five are not suitable as drugs anymore. Thus, the ratio of hydrophilic to hydrophobic side chains was chosen as 2:3 for P322-P3325 and 1:2 for P326-329. *N*lys (R_1) and *N*phe (R_3) were used to represent natural amino acids lysine and phenylalanine. *N*pcb (R_4) was included as a highly lipophilic side chains and 4-butlyamine (*N*but, R_7) and 7-heptlyamine (*N*hep, R_8) were incorporated as amphipatic and nonpolar side chains.

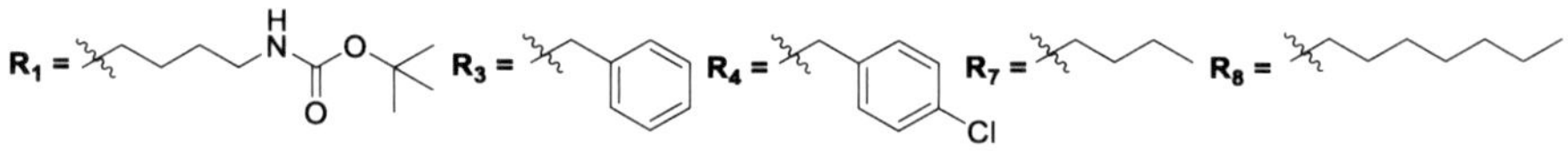

Figure 23: Amines used for synthesis of peptoid library 3: tert-butyl (4-aminobutyl)carbamate (R_1), benzylamine (R_3) and 4-chlorobenzylamine (R_4), butylamine (R_7) and heptylamine (R_8).

An overview of peptoid sequences and their respective molecular weight is shown in table 5.

Table 5: Overview of peptoids in library 3. Peptoid numbers, sequences and respective molecular weights (g/mol) are shown. Success of synthesis was verified by MALDI-TOF mass spectrometry and identified peptoids are marked with "x".

Peptoid	Sequence	Molecular weight (g/mol)	Mass found (x)
322	*N*lys-*N*phe-*N*phe-*N*lys-*N*phe-*N*phe-*N*lys-*N*phe-*N*phe-*N*lys	1836.37	x
323	*N*phe-*N*phe-*N*phe-*N*lys-*N*lys-*N*lys-*N*lys-*N*phe-*N*phe-*N*phe	1836.37	x
324	*N*phe-*N*phe-*N*hep-*N*lys-*N*lys-*N*lys-*N*lys-*N*hep-*N*phe-*N*phe	1852.50	x
325	*N*phe-*N*phe-*N*but-*N*lys-*N*lys-*N*lys-*N*lys-*N*but-*N*phe-*N*phe	1768.34	x

326	Nlys-Nphe-Nphe-Nlys-Nphe-Nphe-Nlys-Nphe-Nphe-Nlys-Npcb-Npcb	2199.61	x
327	Nphe-Nphe-Nphe-Nlys-Nlys-Nlys-Nlys-Nphe-Nphe-Nphe-Npcb-Npcb	2199.61	x
328	Nphe-Nphe-Nhep-Nlys-Nlys-Nlys-Nlys-Nhep-Nphe-Nphe-Npcb-Npcb	2215.74	x
329	Nphe-Nphe-Nbut-Nlys-Nlys-Nlys-Nlys-Nbut-Nphe-Nphe-Npcb-Npcb	2131.58	x

All peptoids could be successfully identified by their molecular weight with MALDI-TOF mass spectrometry and were subsequently purified by HPLC. Analysis of cell penetration abilities was done by confocal microscopy by incubation 10 μM peptoid in 1.5 x 10^4 HeLa cells for 24 h and co-stained mitochondria, with 125 nM MitoTracker® Green, and cell nuclei, with 2 μg/ml Hoechst 33342. Cellular uptake for P338-P344 is shown in figure 24, displaying strong uptake of all analyzed peptoids. Even though, CPPos in library 3 displayed different lipophilicities, all peptoids were accumulating in endosomes. Endocytotic uptake could be triggered by Nlys, which was present in each peptoid or due to increased size of peptoids, hindering them in direct penetration of the plasma membrane. Differences between P322-P329 could not be observed. Even though decameric and dodecameric CPPos displayed strong uptake, their suitability as molecular transporters is questionable as endocytotic uptake is not desirable for the transport of cargo into cells. Most therapeutic molecules have their target in the cytoplasm or nuclei of cells. However, if uptake takes place by endocytosis, escape of endosomes into the cytoplasm is hard to achieve.

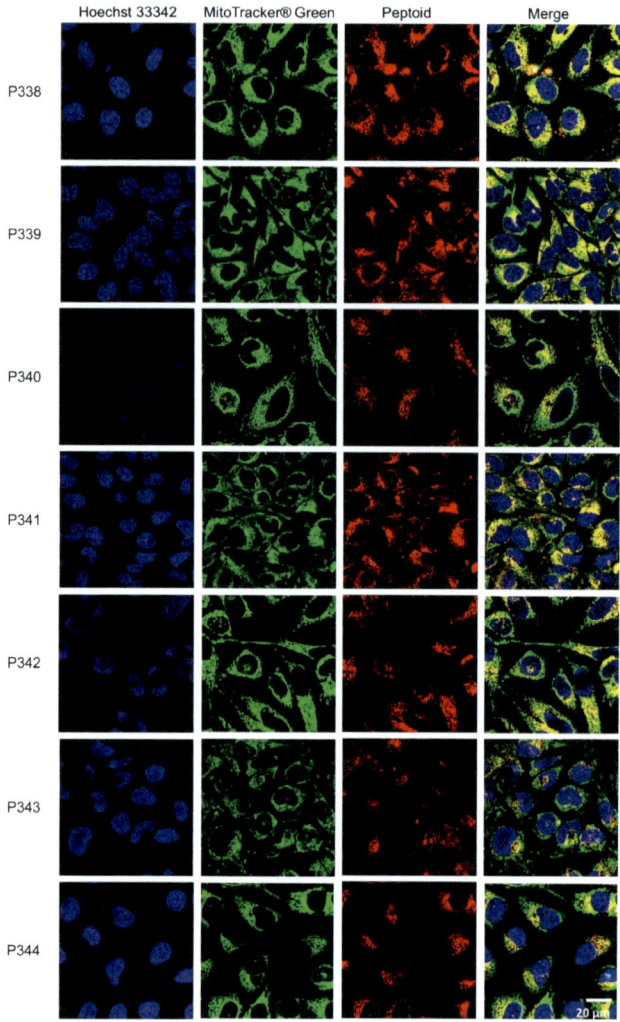

Figure 24: Cellular uptake of P338-P344 in HeLa cells. 1.5 x 10⁴ cells were treated with 10 µM of P338-P344 for 24 h at 37 °C. For co-staining of mitochondria and nuclei cells were treated with 125 nM MitoTracker® Green and Hoechst 33342 (2 µg/ml). Intracellular accumulation of the peptoids was detected with fluorescence confocal microscopy (Leica TCS-SPE, Objective: ACS APO 63x/1.30 OIL). 1. Hoechst 33342, Ex.: 405 nm, Em.: 417-468 nm, 2. MitoTracker® Green Ex.: 488 nm, Em.: 499-552 nm 3. Peptoid, Ex.: 561 nm, Em.: 593-696 nm 4. Merge, Scale bar: 20 µm

3.1.4. Peptoid library 4

Recently cyclization of peptides and peptidomimetics has become a promising method to increase stability and pharmakinetic properties. For example poor bioavailibility of peptides could be improved after cyclization due to hydrophobic side chains, creating a lipophilic surface of the peptide, sheltering cleavable amide bonds in the cycle and simultaneously improve membrane penetration [125]. Furthermore, flexibility of peptides and peptidomimetics is decreased after cyclization and therefore toxicity *in vivo* can be lower due to limitied possible unspecific interactions with receptors [126, 127]. It has been shown that cyclic peptides are promising molecules in the development and screening for new therapeutic agents with bride varities of applications [128-130]. Hence, cyclization can also be an interesting tool in the development of transporter molecules for drug delivery. Lately, many possibilities for cyclization of peptoids have been published. However, many synthesis approaches are not suitable for synthesis of large libraries, for example head-to -tail cyclizations after cleavage from the resin in liquid phase [131-135]. Side-chain cyclization is a promising approach for creation of macrocyclic peptoids on solid-phase and allows following coupling of fluorescent dyes and recently, Kwon et al. was able to synthesis a cyclic one-bead-one-compund peptoid library [136]. As one-bead-one-compound libraries, due to low peptoid concentrations, are not suitable for toxicity studies and *in vivo* anaysis it was investigated if synthesis of cyclic peptoid libraries is possible in IRORI MinsKans. Cyclization was done according to synthesis published by Holub et al. by copper catalyzed 1,-3- alkyne-azide cycloaddition of peptoid side chains [137]. As a *proof of principle* 15 hexameric peptoids with different hydrophobic and hydrophilic side chains were synthezised. For each peptoid alkyne side chain was placed on position two, while azide side chain was placed in position five. Cyclization of hexameric peptoids could be achieved by 1,3- dipolar cycloaddition, on solid phase, by adding copper(I) iodide, DIPEA and ascorbic acid. Subsequently, the resin was washed several times and pepoids were coupled to rhodamine B. Finally, cleavage was done with 95% TFA and success of synthesis was controlled by LC/MS. Cyclization, rhodamine B coupling and cleavage for heaxameric peptoids is shown in scheme 3.

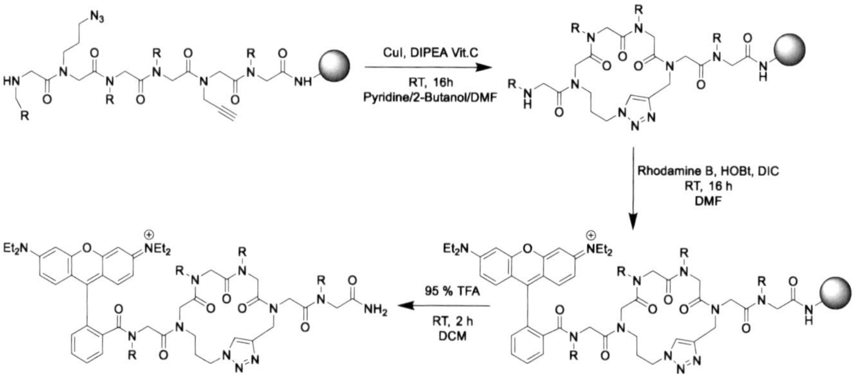

Scheme 4: Cyclization of peptoids containing Nprg (Position 2) and N4az (Position 5) by CuAAc. Subsequently, peptoids are coupled to rhodamine B and finally cleaved from the resin.

Furthermore, linear counterparts were synthesized to controll success of cyclization. Even though cyclic peptoids display the same molecular weight as their linear counterparts, HPLC retention time is different. Cyclic peptoids are more hydrophilic and therefore have earlier retention times. 14 out of 15 peptoids were sucessfully synthezised and could be identified by their mass with LC/MS and MALDI-TOF (table 6).

Table 6: Overview of cyclic peptoids and their linear counterparts in library 4. Peptoid numbers, sequences, respective molecular weights (g/mol) and retention times (min) are shown. Success of synthesis was verified by MALDI-TOF mass spectrometry and identified peptoids are marked with "x".

Peptoid	Sequence	Molecular weight [g/mol] (found = x)	Retention time [min]	Linear	Retention time [min]
Cyclo 1	Nphe-Nprg-Nphe-Nphe-N4az-Nphe	1266.54 (x)	31:00	x	32:30
Cyclo 2	Nphe-Nprg-Npob-Npob-N4az-Nphe	1298.54 (x)	27:00	x	29:15
Cyclo 3	Nphe-Nprg-Nlys-Nlys-N4az-Nphe	1228.53 (x)	21:50	x	22:45
Cyclo 4	Nphe-Nprg-Npob-Npob-N4az-Npob	1314.53 (x)	30:30	x	32:25
Cyclo 5	Nphe-Nprg-Npfb-Npfb-N4az-Npob	1318.52 (x)	-	x	-
Cyclo 6	Nphe-Nprg-Nphe-Nphe-N4az-Npfb	1284.53 (x)	32:50	x	34:00
Cyclo 7	Nphe-Nprg-Npfb-Nlys-N4az-Npfb	1283.52 (x)	26:20	x	28:15
Cyclo 8	Nhex-Nprg-Nhex-Nhex-N4az-Nhex	1242.69 (x)	28:00	-	-
Cyclo 9	Nhex-Nprg-Npcb-Npbf-N4az-Nhex	1307.04 (x)	26:00	-	-
Cyclo 10	Nhex-Nprg-Nhex-Nlys-N4az-Nlys	1216.61 (x)	23:00	-	-

Cyclo 11	Npcb-Nprg-Npcb-Npcb-N4az-Npcb	1404.31 (x)	25:00	-	-
Cyclo 12	Npcb-Nprg-Npbf-Npbf-N4az-Npcb	1371.40 (-)	-	-	-
Cyclo 13	Npcb-Nprg-Npcb-Npbf-N4az-Npbf	1371.40 (x)	27:00	-	-
Cyclo 14	Npbf-Nprg-Npbf-Npbf-N4az-Npbf	1338.50 (x)	32:00	-	
Cyclo 15	Npbf-Nprg-Nhex-Nlys-N4az-Npbf	1277.55 (x)	31:00	-	-

All cyclic peptoids and their linear counterparts were purifed by HPLC and cyclic peptoids displayed earlier retention times, as already reported by Holub et al.. Dimerized peptoids, which would be more lipophilic as monomeric cycles and therefore have later retention times, could not be identified. 12 cyclic peptoids were sucessfully purified and isolated with exception of cyclo 5 and cyclo 13, which could not be purified due to low synthesis yield. Subsequently, cellular uptake and localization of peptoids was analyzed in 1.5 x 10^4 HeLa cells after 24 h incubation time. Cell nuclei and mitochondria were visualized by co-staining with MitoTracker®Green and Hoechst 33342 and comparison of cyclic peptoids with linear peptoids is shown in figure 25 and figure 26. Interestingly, intracellular localization of peptoids changes after cyclization. Peptoid 1 and peptoid 6 are highly hydrophobic, consisting only of Nphe and Npfb, and linear peptoids accumulated, as expected, in mitochondria. In contrast, cyclo 1 and cyclo 6 displayed mainly uptake by endocytosis. Peptoid 3 and peptoid 7 are more hydrophilic, as they contain Nlys. While linear 3 and linear 7 accumulated in endosomes, cyclo 3 and cyclo 7 were located not only in endosomes but also in mitochondria. For peptoid 2 and 4, containing Nphe and Npob, no difference between linear and cyclic peptoid was visible, as both are located in endosomes and mitochondria.

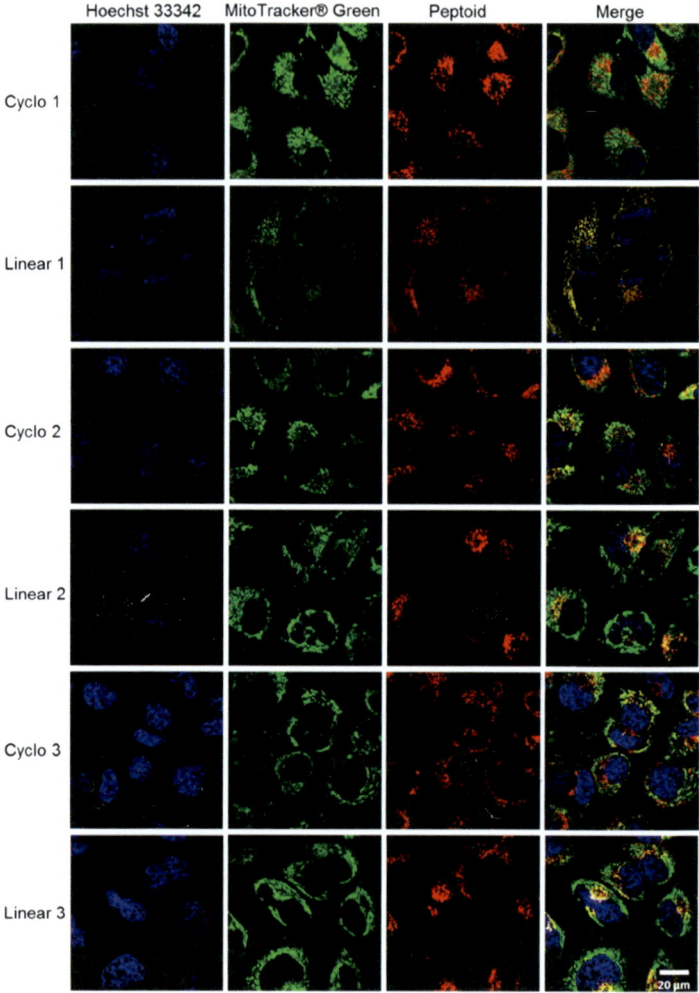

Figure 25: Cellular uptake of cyclic peptoids in comparison to linear peptoids in HeLa cells. 1.5 x 10^4 cells were treated with 10 µM Cyclo 1-3 and Linear 1-3 for 24 h at 37 °C. For co-staining of mitochondria and nuclei cells were treated with 125 nM MitoTracker® Green and Hoechst 33342 (2 µg/ml). Intracellular accumulation of the peptoids was detected with fluorescence confocal microscopy (Zeiss LSM800, Objective: Plan-Apochromat 63x/1.40 Oil DIC). 1. Hoechst 33342, Ex.: 405 nm, Em.: 417-468 nm, 2. MitoTracker® Green Ex.: 488 nm, Em.: 499-552 nm 3. Peptoid, Ex.: 561 nm, Em.: 593-696 nm 4. Merge, Scale bar: 20 µm

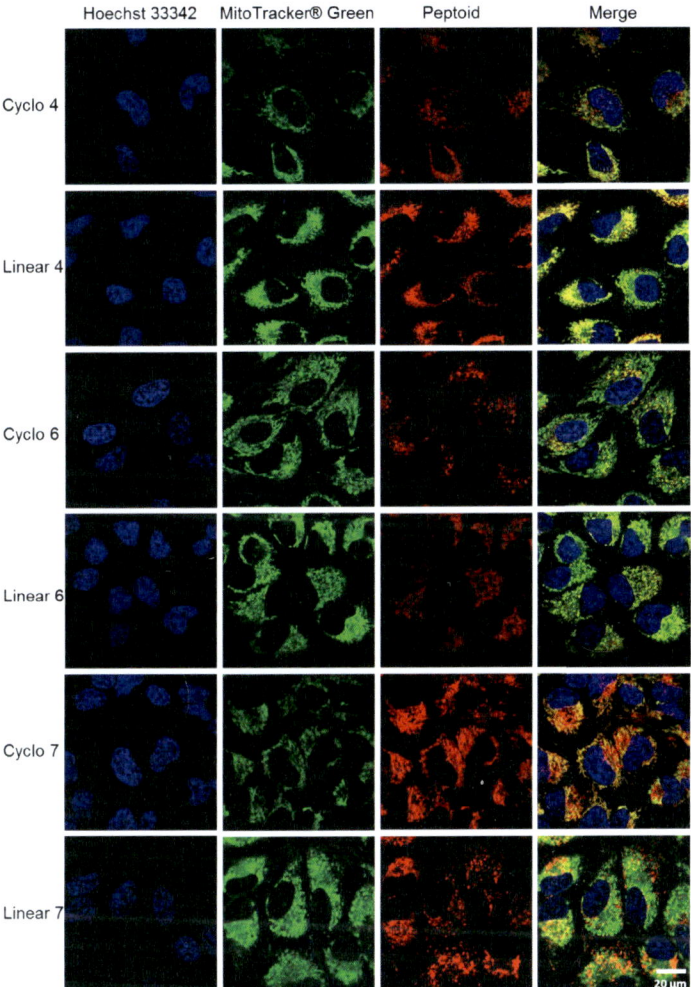

Figure 26: Cellular uptake of cyclic peptides in comparison to linear peptoids in HeLa cells. 1.5 x 10^4 cells were treated with 10 µM Cyclo 4-7 and Linear 4-7 for 24 h at 37 °C. For co-staining of mitochondria and nuclei cells were treated with 125 nM MitoTracker® Green and Hoechst 33342 (2 µg/ml). Intracellular accumulation of the peptoids was detected with fluorescence confocal microscopy (Zeiss LSM800, Objective: Plan-Apochromat 63x/1.40 Oil DIC). 1. Hoechst 33342, Ex.: 405 nm, Em.: 417-468 nm, 2. MitoTracker® Green Ex.: 488 nm, Em.: 499-552 nm 3. Peptoid, Ex.: 561 nm, Em.: 593-696 nm 4. Merge, Scale bar: 20 µm

For the remaining cyclic peptoids, cyclo 8-11 and cyclo 14 and 15 the linear counterpart was not synthesized. Uptake of those peptoids is shown in figure 27, confirming that highly lipophilic and aromatic cyclic peptoids display endosomal uptake. Cyclo 11 and cyclo 14 consits only of Npbc (11) and Nbpf (14) and for linear peptoids mitochondrial uptake would be expected. However, as shown in figure 27, cyclo 11 and cyclo 14 displayed mainly uptake by endocytosis. Lipophilic peptoid cyclo 9 contains not only aromatic side chains (Nbpf, Npbc) but also aliphatic side chains (Nhex) and was accumulating in both organelles, endosomes and mitochondria. Hydrophilic peptoids cyclo 10 and cyclo 15 accumulated mainly in endosomes and low concentrations of cyclo 15 were also found in mitochondria. A possible explanation for endosomal uptake of highly lipophilic, aromatic cyclic peptoids could be π-π stacking between aromatic side chains within the peptoid and decreased flexibility of the peptoid chain, hindering direct penetration and increasing uptake by endocytosis.

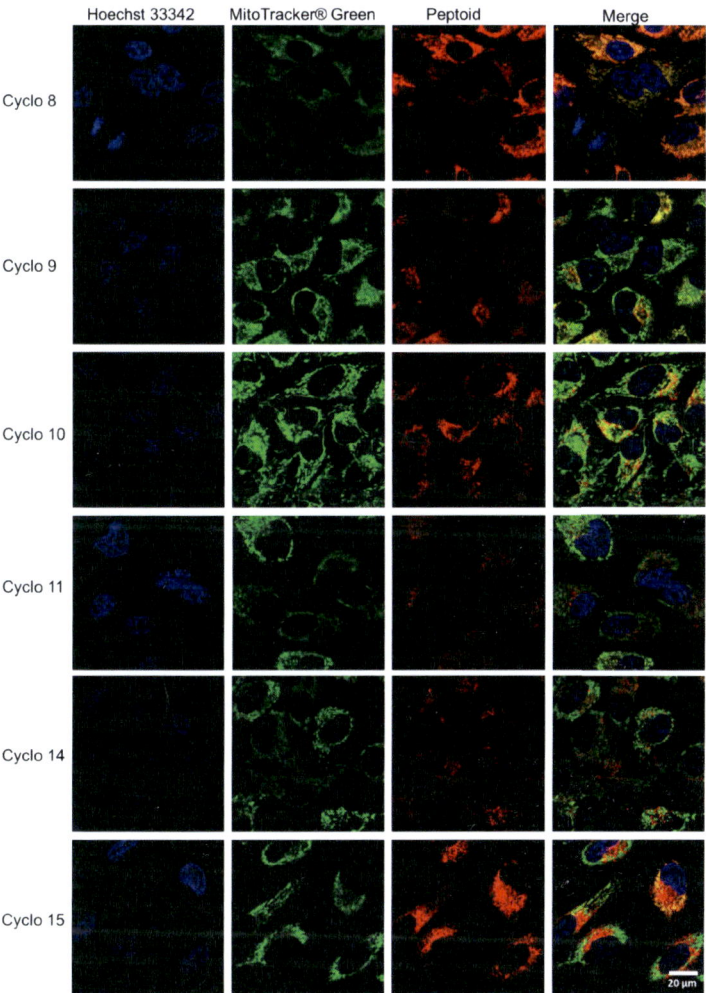

Figure 27: Cellular uptake of cyclic peptoids in HeLa cells. 1.5 x 10⁴ cells were treated with 10 µM Cyclo 8-15 for 24 h at 37 °C. For co-staining of mitochondria and nuclei cells were treated with 125 nM MitoTracker® Green and Hoechst 33342 (2 µg/ml). Intracellular accumulation of the peptoids was detected with fluorescence confocal microscopy (Zeiss LSM800, Objective: Plan-Apochromat 63x/1.40 Oil DIC). 1. Hoechst 33342, Ex.: 405 nm, Em.: 417-468 nm, 2. MitoTracker® Green Ex.: 488 nm, Em.: 499-552 nm 3. Peptoid, Ex.: 561 nm, Em.: 593-696 nm 4. Merge, Scale bar: 20 µm

3.1.4.1. Cytotoxicity of cyclic peptoids

Cytotoxicity of cyclic peptoids was compared to their linear counterparts for cyclo 1, cyclo 2 and cyclo 3 using an MTT assay to determine the viability of cells. Therefore, 0.75×10^4 HeLa cells/well were incubated with the respective peptoid (10, 30 and 60 µM) for 72 h. The evaluation of the MTT assay is shown in figure 28, displaying significant differences for the toxicity of cyclic peptoids in comparison to linear peptoids.

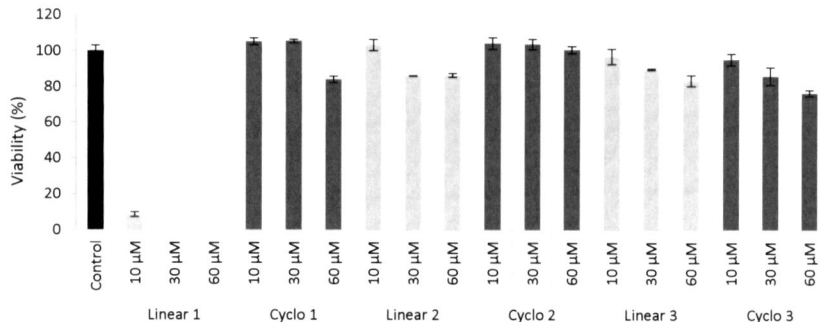

Figure 28 : Cytotoxicity of cyclic peptoids and their linear counterpart in HeLa cells. Peptoids (Linear 1, Cyclo 1, Linear 2, Cyclo 2, Linear 3, Cyclo 3) were tested in 10, 30 and 60 µM concentration for 72 h and viability was determined using the MTT assay.

The greatest difference in toxicity between the cyclic and linear peptoid counterpart was determined for linear 1 / cyclo 1, a hydrophobic peptoid containing Nprg and N4az once and Nphe four times. While linear 1 was highly toxic ($LD_{50} < 10$ µM), even for 10 µM, cyclo 1 displayed low toxicity in Hela cells and LD_{50} value could not be determined as it was beyond the tested concentrations. Differences in cytotoxic effects of the peptoids can be explained by differences in cellular uptake of those peptoids. For linear 1 strong mitochondrial uptake was found, which can led to enhanced toxicity due to disturbance of cellular functions as described in chapter 3.1.1.3.. In contrast, cyclo 1 was taken up by endocytosis and accumulation of peptoids in endosomes is usually less toxic to cells, as peptoids cannot interfere in cellular functions. The differences between linear 2 and cyclo 2, a peptoid containing Nphe, Nprg, Npob and N4az, were smaller, but still significant for 30 µM and 60 µM. For those peptoids no visual differences in cellular uptake were found, however, while cyclo 2 displayed no toxicity to cells, low toxic effects were found for linear 2. Differences for linear 3 and cyclo 3, a hydrophilic peptoid containing Nphe, Nlys, Nprg and N4az, are small for high concentrations

(60 µM) and no differences were found for low concentrations (10 µM and 30 µM). For 60 µM cyclo 3 was slightly more toxic compared to linear 3, which was properly triggered by the mechanism of cellular uptake, which was found to be endosomal and mitochondrial, for cyclo 3. These findings confirm, that cyclization of peptoids changes their toxicity to cells, however, with respect to their composition of side chains they can also display enhanced toxicity.

3.2.   Anticancer Peptoids

High-throughput screening and toxicity tests in HeLa cells of CPPo library 1 displayed several MPPos, which are suitable as molecular transporters due to their high affinity to mitochondria, strong cellular uptake and low toxicity to cells. However, some candidates showed increased toxicity to cells (chapter 3.1.1.3). Hence, those peptoids might not be suitable as carrier molecules for drugs. However, it has been shown that CPPos are also promising drugs for anticancer treatments [79]. Besides surgery and radiation therapy, chemotherapy is one of the most common treatments for cancer patients. At the moment, a wide amount of cytotoxic chemotherapeutic agents is available on the pharmaceutical market, including anthracyclines, peptides, nucleotide analogs and platinum-based agents, interfering with the basic machinery of the cell [138-142]. However, one huge drawback of common anticancer drugs is their unspecificity, leading to side effects, such as pain, organ damage and anemia [143]. Another problem in chemotherapy is the resistance of cancer cells to common drugs, due to overexpression of the *mdr* gene, encoding for the P-glycoprotein, an efflux pump discharging a broad spectrum of molecules [144, 145]. Furthermore, some anticancer drugs are known to be cancerogenic and can induce secondary cancer, such as acute myelogenous leukaemia [146-148]. Recently a wide range of cell penetrating peptides has been tested as potential cytotoxic cancer drugs, however, their suitability is limited, due to their rapid degradation *in vivo* [149-151]. 18 MPPos, shown in table 7, with an LD_{50} value less than or equal to 40 µM (HeLa cells), were tested as potential mitochondria-targeting anticancer agents. The most occurring side chain was *N*pcb (51%), which was present in each peptoid at least once and only one peptoid contained *N*lys (P182).

Table 7: Overview of peptoids which were investigated as potential anticancer peptoids. Peptoid numbers, sequences and LD_{50} values (72 h, HeLa) are shown.

Peptoid	Sequence	LD_{50} HeLa
P54	Nprg-Nphe-Npcb-Npcb-RhodB	~ 16 µM
P82	Npcb-Npcb-Nprg-Npcb-RhodB	~ 25 µM
P85	Npcb-Npcb-Npcb-Nprg-RhodB	~ 40 µM
P86	Npcb-Npcb-Npcb-Npcb-RhodB	~ 20 µM
P88	Npcb-Npcb-Npcb-Nphe-RhodB	~ 20 µM
P93	Npcb-Npcb-Nphe-Nprg-RhodB	~ 30 µM
P96	Npcb-Npcb-Nphe-Nphe-RhodB	~ 27 µM
P114	Npcb-Nphe-Nprg-Npcb-RhodB	~ 22.5 µM
P118	Npcb-Nphe-Npcb-Npcb-RhodB	~ 40 µM
P120	Npcb-Nphe-Npcb-Nphe-RhodB	~ 25 µM
P128	Npcb-Nphe-Nphe-Nphe-RhodB	~ 20 µM
P182	Nlys-Nphe-Npcb-Npcb-RhodB	~ 15 µM
P194	Nphe-Nprg-Nprg-Npcb-RhodB	~ 30 µM
P200	Nphe-Nprg-Npcb-Nphe-RhodB	~ 30 µM
P214	Nphe-Npcb-Npcb-Npcb-RhodB	~ 13 µM
P216	Nphe-Npcb-Npcb-Nphe-RhodB	~ 13 µM
P221	Nphe-Npcb-Nphe-Nprg-RhodB	~ 30 µM
P246	Nphe-Nphe-Npcb-Nprg-RhodB	~ 30 µM

Cytotoxicity of eight peptoids (P54, P82, P85, P96, P114, P118, P182 and P216) was also analyzed in two other cancer cell lines, HepG2 (human liver cancer cells) and MCF-7 (human breast cancer cells). Furthermore, selective toxicity of peptoids to cancer cells was analyzed by testing their cytotoxicity in HUVEC (human umbilical vein endothelial cells) and NHDF (normal human dermal fibroblasts). Cytotoxicity was determined for 5, 10, 20 and 40 µM peptoid solution, for 72 h, using the MTT assay. Cell viability and LD_{50} values for tested peptoids are shown in figure 29, displaying that peptoids were not only cytotoxic for HeLa cells but also for HepG2 and MCF-7 cells. A comparison of their cytotoxicity in different cancer cell lines showed high cytotoxicity of peptoids especially in HepG2 cells. The most toxic peptoid was P216 with an LD_{50} value between 8.5 and 14 µM. However, peptoids also displayed increased cytotoxicity to endothelial cells and no significant difference between cancer cells and HUVEC cells could be determined. A possible explanation for increased cytotoxicity of peptoids in HUVEC cells is the unnatural cultivation method of HUVEC cells in a 2D monolayer. In this flat environment they might be more sensitive to chemotherapeutics. Cytotoxicity to NHDF cells was significant lower compared to cancer cells lines, which confirms the findings of Huang et al., that peptoids show increased toxicity to cancer cells compared to primary dermal fibroblasts [79]. For P85, P118 and P182 LD_{50} values for NHDF cells could not be determined, as they had no toxic effects on cells for the tested concentrations. P82 and P96 were significant less toxic to NDHF cells for low concentrations ($\leq$20 µM) and displayed higher LD_{50} values compared to cancer cell lines. Only for the highest tested concentration (40 µM), they were cytotoxic in NDHF as well. For peptoids P54, P114 and P216 the difference between cancer cells and fibroblasts was rather low. Comparing the structures of the peptoids and their cytotoxicity in cancer cells compared to dermal fibroblasts showed that peptoids with the lowest toxic effects on fibroblasts were rich in Npcb (67%), while the amount of Nphe (17%) was rather low. The most occurring side chain for the peptoids which displayed increased toxicity to NHDF cells was still Npcb (50%), however, those peptoids were also rich in Nphe (33.3%). The ratio between aromatic and non-aromatic side chains was similar for both.

Table 8: LD_{50} values (72 h) for P54, P82, P85, P88, P96, P114, P118, P182, P214 and P216 in HeLa, HepG2, MCF-7, HUVEC and NHDF cells.

Peptoid	LD_{50} HeLa	LD_{50} HepG2	LD_{50} MCF-7	LD_{50} HUVEC	LD_{50} NHDF
P54	~ 16 µM	~ 16 µM	~ 13 µM	~ 13 µM	~ 18 µM
P82	~ 25 µM	~ 17 µM	~ 20 µM	~ 17.5 µM	~ 29 µM
P85	~ 40 µM	~ 17 µM	~ 27.5 µM	~ 26 µM	> 40 µM
P88	~ 20 µM	~ 15 µM	~17.5 µM	~ 27.5 µM	> 40 µM
P96	~ 27 µM	~ 15 µM	~ 20 µM	~ 22.5 µM	~ 32 µM
P114	~ 22.5 µM	~ 10 µM	~ 17,5 µM	~ 17 µM	~ 16 µM
P118	~ 40 µM	~ 14 µM	~ 40 µM	~ 28 µM	>40 µM
P182	~ 15 µM	~ 18 µM	~ 38 µM	>40 µM	>40 µM
P214	~ 13 µM	~17.5 µM	~ 10 µM	~ 24 µM	~ 32 µM
P216	~ 13 µM	~ 8.5 µM	~ 14 µM	~ 9 µM	~ 14 µM

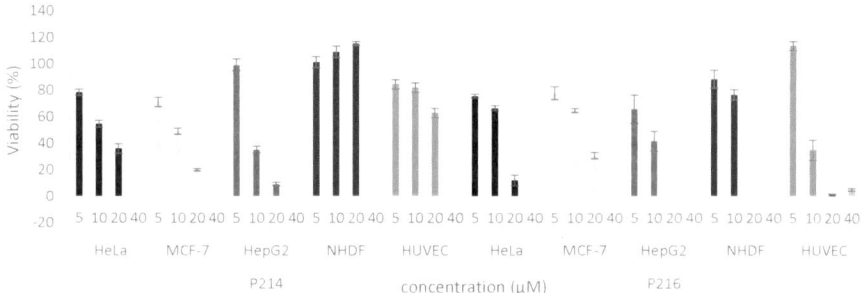

Figure 29 : Cytotoxicity of potential anticancer peptoids in different primary (HUVEC, NHDF) and cancer cells (HeLa, HepG2, MCF-7). P54, P82, P85, P88 P96, P114, P118, P182, P214 and P216 were tested in 5, 10, 20 and 40 µM concentration for 72 h and viability was determined using the MTT assay.

For a better understanding of the cytotoxic effects of peptoids in cancer cells the pathway of cell death was analyzed. Apoptosis, the normal, programmed cell death, is usually preferred for anticancer drugs, as this form of cell death leads to nuclear fragmentation, without damage of cellular membranes and therefore no release of DNA and possible damage of neighboring tissue. This form of cell death can be identified after staining cell nuclei with Hoechst 33342 and microscopic visualizing of reduction and fragmentation of the nuclei. In comparison to that necrosis leads to a disruption of membranes and liberation of potentially cytotoxic DNA, which might affect the healthy surrounding cells. Typically, necrotic cells are identified by staining with propidium iodide, a fluorescent intercalating agent, which is not able to cross cellular membranes of viable cells. Hence, it can only stain necrotic cells, due to their damaged cell membranes. However, the excitation and emission of propidium iodide (Ex.$_{max}$: 535 nm; Em.$_{max}$: 617 nm) overlaps with excitation and emission of rhodamine B labeled peptoids (Ex.$_{max}$: 553 nm; Em.$_{max}$: 627 nm). As an alternative to propidium iodide, cells were stained with trypan blue, which cannot cross cellular membranes as well and is not fluorescent. Uptake of trypan blue in necrotic cells can be visualized in brightfield images. 1.5×10^4 HeLa and MCF-7 cells were incubated with peptoid solution (at the respective LD$_{50}$ concentration) overnight and afterwards washed with DPBS and stained with Hoechst 33342 and trypan blue. Cells were analyzed by fluorescence confocal microscopy and necrotic and apoptotic cells were quantified afterwards. Cell death pathway was analyzed for the 18 potential anticancer MPPos for HeLa and MCF-7 cells and proportions are shown in figure 30 for both cell lines.

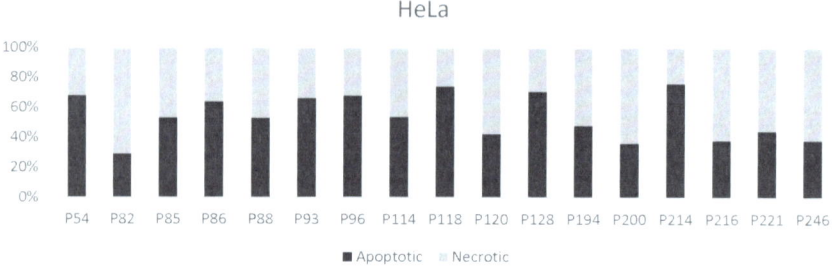

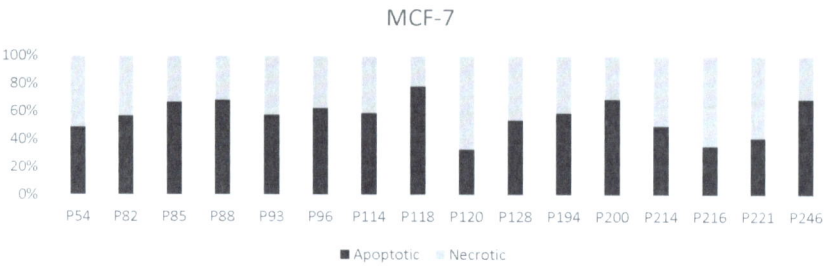

Figure 30 : Pathway of cell death for 1.5 x 10^4 HeLa and MCF-7 cells incubated with potential anticancer peptoid (LD$_{50}$ concentration) for 24 h. Necrotic cells were identified by staining with trypan blue and apoptotic cells by staining with Hoechst 33342 and identification of fragmentation of nuclei.

Analysis of cell death showed, that the peptoids had different effects on the cells, while some peptoids led mainly to apoptotic cell death others induced necrosis. Furthermore, effects differ for different tested cell lines. In HeLa cells P54, P86, P88, P93, P96, P114, P118, P128 and P214 led predominantly to apoptotic cell death (>60% apoptotic cells). In MCF-7 cells P85, P88, P96, P118, P200, P246 were found to induce apoptotic cell death. Hence, only three peptoids (P88, P96 and P118) tend to induce apoptotic cell death in both treated cell lines. All of them consist only of Npcb and Nphe. Potential anticancer MPPos were also screened for induction of reactive oxygen species (ROS). It was found that the level of ROS in cancer cells is increased compared to healthy cells, amongst others, due to their high metabolic activity and mitochondrial dysfunction [152]. Additionally, it is supposed, that low concentrations of ROS enhance development and progression of tumors. However, cancer cells also display expression of antioxidant proteins to detoxify from ROS and it can be assumed that only certain concentrations of intracellular ROS are appropriate for cellular function of cancer cells [153, 154]. High concentrations of intracellular ROS can stop the cell cycle and lead to apoptotic cell death [155]. Lately it has been shown, that ROS inducing drugs are effectively killing selective cancer cells without damaging the surrounding healthy tissue and therefore production of ROS is a promising feature for anticancer drugs [156]. Intracellular concentrations of ROS can be detected by staining cells with the cell-permeant profluorescent 2',7'-dichlorodihydrofluorescein diacetate (H$_2$DCFDA). After penetrating cells, cellular esterases cleave the acetate group revealing 2',7'-dichlordihydrofluorescein (H$_2$DCF), which is still profluorescent. In the presence of ROS H$_2$DCF gets oxidized to the highly fluorescent 2',7'-dichlorfluorescein (DCF) [157].

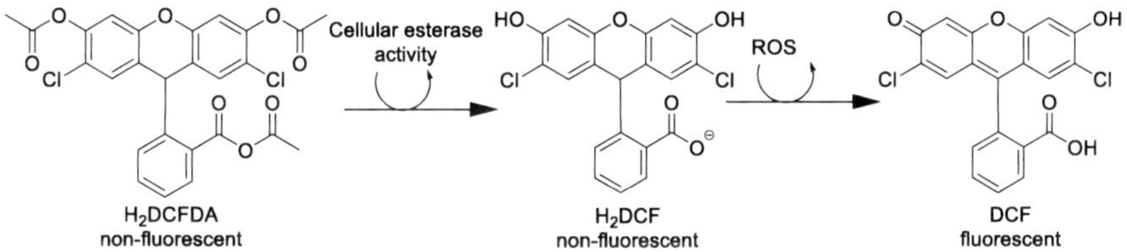

Scheme 5: Conversion of H$_2$DCFDA (non-fluorescent), after cleavage of acetate group by intracellular esterases, to H$_2$DCF (non-fluorescent). Subsequently, oxidation to fluorescent DCF in the presence of reactive oxygen species.

To investigate ROS production of peptoids 1.5 x 10^4 HeLa cells were incubated for 6 h with 30 µM peptoid solution and after 5 h incubation 5 µM H$_2$DCFDA was added and cells were incubated for one further hour. Finally, cells were washed with DPBS and cell nuclei were stained with Hoechst 33342. Cells incubated without peptoids served as negative control, while H$_2$O$_2$ was used as positive control. Subsequently, cells were analyzed by confocal microscopy and DCF was detected by excitation with 488 nm and emission between 517 and 527 nm.

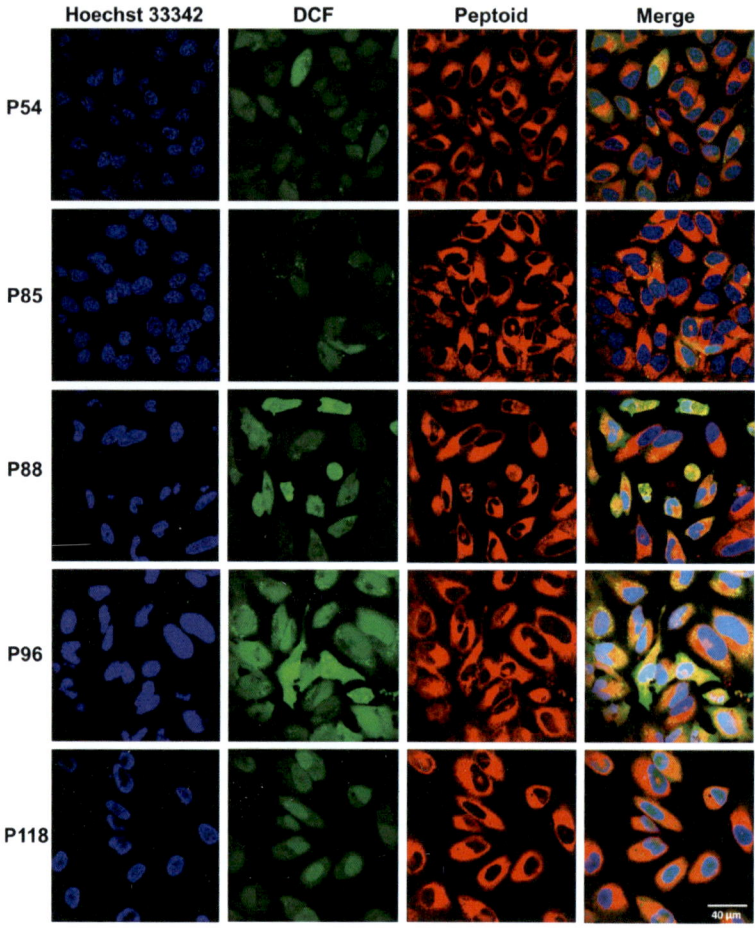

Figure 31: Detection of intracellular reactive oxygen species in HeLa cells induced by peptoids. 1.5 x 10^4 cells were treated with 30 μM P54, P85, P88, P96 and P118 for 6 h at 37 °C. Reactive oxygen species were detected by incubation with H$_2$DCFDA. For co-staining of nuclei cells were treated with Hoechst 33342 (2 μg/ml). Intracellular generation of ROS was investigated by fluorescence confocal microscopy (Leica TCS-SPE, Objective: ACS APO 63x/1.30 OIL). 1. Hoechst 33342, Ex.: 405 nm, Em.: 417-468 nm, 2. DCF Ex.: 488 nm, Em.: 510-520 nm 3. Peptoid, Ex.: 561 nm, Em.: 593-696 nm 4. Merge, Scale bar: 40 μm

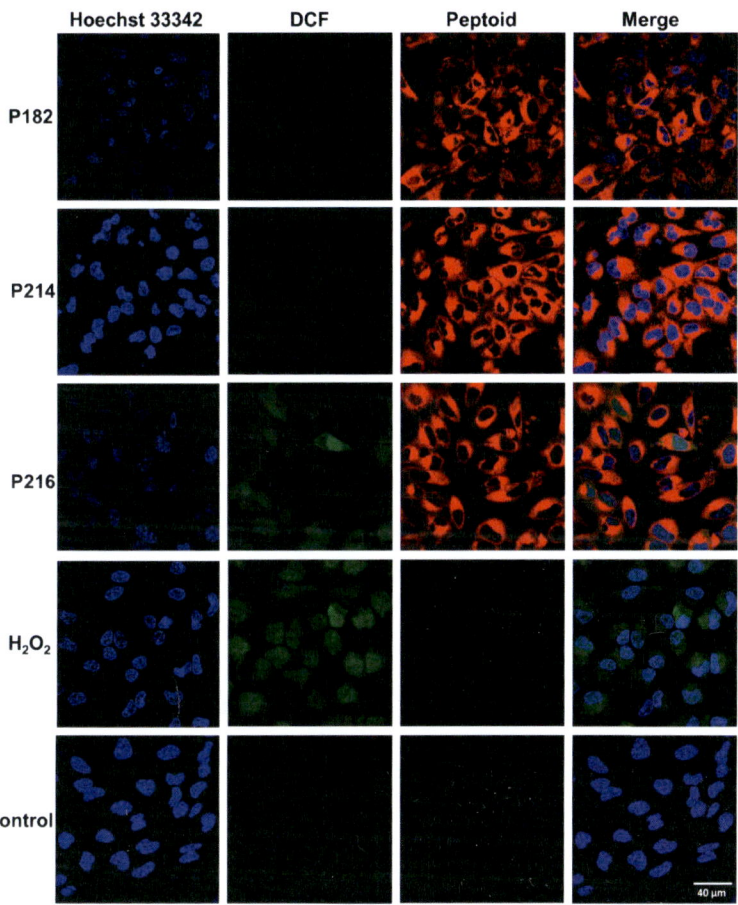

Figure 32: Detection of intracellular reactive oxygen species in HeLa cells induced by peptoids. 1.5 x 10⁴ cells were treated with 30 μM P182, P214 and P216 for 6 h at 37 °C. Controls were done by incubating cells with H_2O_2 and untreated cells served as negative control. Reactive oxygen species were detected by incubation with H₂DCFDA. For co-staining of nuclei cells were treated with Hoechst 33342 (2 μg/ml). Intracellular generation of ROS was investigated by fluorescence confocal microscopy (Leica TCS-SPE, Objective: ACS APO 63x/1.30 OIL). 1. Hoechst 33342, Ex.: 405 nm, Em.: 417-468 nm, 2. DCF Ex.: 488 nm, Em.: 510-520 nm 3. Peptoid, Ex.: 561 nm, Em.: 593-696 nm 4. Merge, Scale bar: 40 μm

Exemplary confocal microscopy images of DCF signals are shown for representative peptoids P54, P85, P88, P96, P118, P182, P214 and P216, as well as negative control (untreated) and positive control (H_2O_2) in figure 31 and figure 32 (P82, P120, P128, P194, P200, P221 and P246: figure 82, appendix). It was found that most peptoids led to an increased intracellular ROS level. Only for P182, the exclusive analyzed peptoid containing Nlys, a DCF signal could not be detected at the investigated concentration.

3.2.1. 3D cell culture

So far, all experiments were done in two-dimensional (2D) cell culture, a suitable method for first studies and preliminary screening of drugs. However, monolayer cell culture might change cell functions and signaling and the architecture and environment of natural occurring tumors cannot be represented [158, 159]. Animal tests are frequently used for *in vivo* analysis of drugs. However, despite the ethical dilemma, animals cannot represent the human tissue. Furthermore, experiments are costly and not suitable for high-throughput screenings of compound libraries [160]. A good alternative and pretesting for animal experiments are tumor spheroids. Cancer cells growing in tumor spheroids, mimicking solid tumors, are more representative for the *in vivo* environment of cancer cells, as cell-cell signaling, physical cell-cell interactions and cellular heterogeneity can be reproduced [161]. Furthermore, it was found that three-dimensional (3D) spheroids, in contrast to monolayer cell culture, are a suitable model to analyze degraded penetration of drugs or even drug resistance occurring *in vivo* [162, 163]. Spheroids can be easily generated by seeding cells on low adhesion plates or in hanging drop plates leading to large amounts of spheroids under reproducible conditions. Hence, spheroids are still suitable to analyze large compound libraries for cellular uptake and toxicity in high-throughput screenings [164]. Many cancer cells can be grown in spheroids, receiving 3D *in vitro* models for several different cancer types. Furthermore, cellular heterogeneity can be achieved by co-culturing other cells, such as endothelia cells or fibroblasts [165, 166]. Here, 4 x 10^3 human melanoma cells (SK-MEL 28) were seeded in 96-well plates filled with 1.5% agarose. After 24 h incubation, cellular growing was enhanced by adding medium. After 72 h incubation and growing of cells, 10 µM peptoid solution was added to the medium surrounding the spheroids. In a first experiment spheroids were incubated with seven peptoids (P80, P82, P85, P88, P96, P118 and P214) for 24 h and uptake of the peptoids was controlled by confocal microscopy. To visualize not only the outer cell layer but also cells in the core of the spheroid, pictures were taken also in z-stacks. Figure 33 shows that representative peptoids, P85, P96 and P118, were located mainly in outer cells, which were in contact with the peptoid solution, however, lower peptoid signal were also detectable in inner cells. Similar results were found for P80, P82, P88, and P214 and fluorescent intensity and penetration depth did not differ for the tested peptoids (figure 80, appendix).

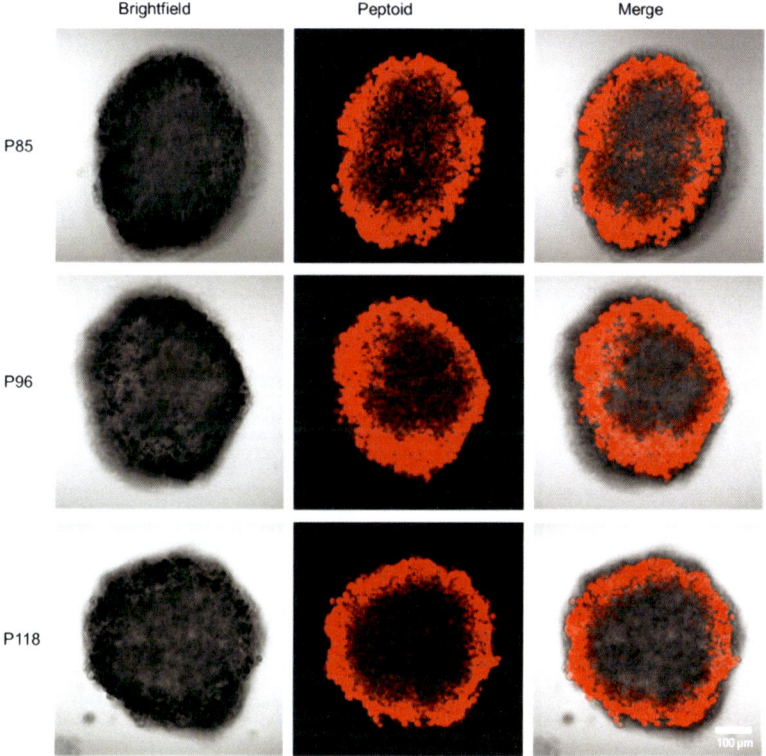

Figure 33: Analysis of intracellular accumulation of peptoids in spheroids of 4 x 10^3 SK-MEL 28 cells. Spheroids were treated with 10 µM P85, P96 and P118 for 24 h at 37 °C. Subsequently spheroids were analyzed with fluorescence confocal microscopy (Zeiss LSM800, Objective: Plan-Apochromat 20x/0.8 M27). 1. Brightfield, 2. Peptoid, Ex.: 561 nm, Em.: 593-696 nm 3. Merge, Scale bar: 100 µm

Furthermore, after treatment with 20 and 40 µM peptoid solution the growth of the spheroids was monitored for five days, providing an insight of the cytotoxic impact of peptoids in 3D cell culture. Therefore, spheroids were generated as described above and after 72 h spheroids were analyzed by microscopy and diameters of spheroids were determined using ImageJ. Impacts of potential anticancer peptoids were analyzed by measuring diameters every 24 h after treatment with 40 µM peptoid solution. Furthermore, impacts of P55 (Nprg-Nphe-Npcb-Nlys-RhodB), an endosomal CPPo, showing no cytotoxic effects in HeLa cells in 2D cell culture, was serving as a negative control peptoid. Microscopy images of spheroids treated with P118 and P214 after 1, 2, 3 and 4 days of treatment, as well as a non-treated control spheroid, are shown in figure 34. While a size increase of the control spheroid was clearly visible especially for the first three days, P118 and P96 led to a strong reduction of spheroid size. For P96 cell-cell-contacts were strongly decreasing after 48 h incubation leading to a broad and loose structure of the spheroid.

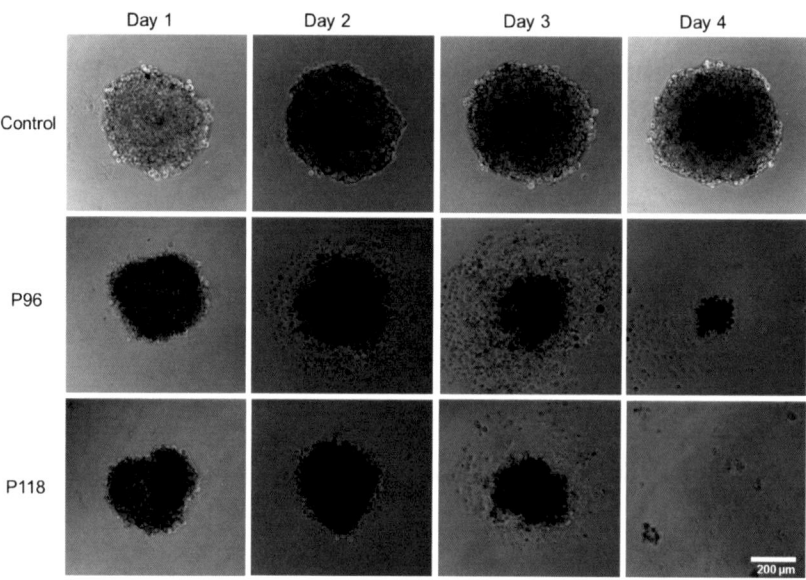

Figure 34: Analysis of the growth of SK-MEL 28 spheroids (4 x 10^3 cells) treated with 40 μM P118 and P214 for 4 days and non-treated control spheroids. Brightfield images were taken with a Leica LED microscope (Leica DMIL LED microscope, HI PLAN 4x/0.10 DRY Objective), Scale bar: 200 μm

These visual findings could be confirmed by measuring the size of spheroids. Figure 35 shows an increasing size of control spheroid and spheroids incubated with control peptoid P55. In contrast, all other peptoids had toxic effects on cells. P214 showed the strongest toxicity revealing complete cell death after three days, approving 2D cell culture results, where P214 showed the lowest LD$_{50}$ value (12.5 μM). P82, P85 and P114 led to complete spheroid damage after four days incubation and most other peptoids induced strong reduction of spheroid size. Only P128 (Npcb-Nphe-Nphe-Nphe-RhodB) and P200 (Nphe-Nprg-Npcb-Nphe-RhodB) were found to have fewer toxic impacts on the spheroid growth, probably due to their low Npcb content. Growth curve of P182 (Nlys-Nphe-Npcb-Npcb-RhodB), the only tested peptoid containing Nlys, is comparable to the control peptoid, displaying no toxicity to cells.

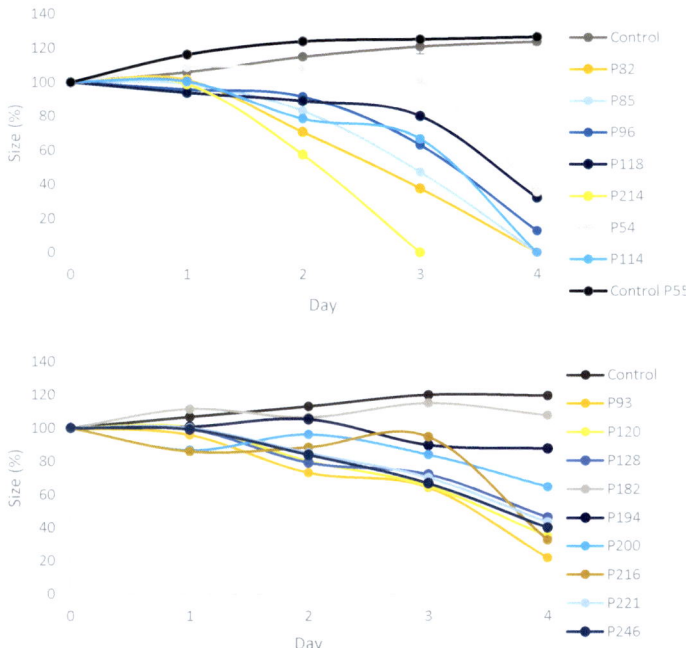

Figure 35: Analysis of the growth of SK-MEL-28 spheroids (4 x 10³ cells) treated with 40 µM of the respective "anticancer" peptoids, control peptoid P55 and non-treated control spheroids. Size of spheroids was measured after 1, 2, 3 and 4 days treatment.

The analysis of spheroids showed the suitability of MPPos as anticancer peptoids and confirmed the cytotoxic effects, which were found in 2D cell culture. Especially peptoids rich in Npcb efficiently stopped cell growth of spheroids and decreased their size intensely within four days.

3.2.1.1. Investigation of peptoids in a microfluidic bioreactor

Cancer cell spheroids are a suitable model to mimic the natural architecture and growth of tumors, including physical, mechanical and cell-cell interactions [161]. However, they cannot be analyzed in connection with blood circulation or in the correct arrangement with different cell types. To overcome this problem, without using animal models, the microfluidic chip system *vasQchip* was developed [167]. This chip system contains a porous microchannel, made from polycarbonate membranes, which can be coated with endothelia cells, and mimics a blood vessel. The surrounding compartment can be used for co-cultivation of different cell types, including primary cells, cancer cells or even cancer cell spheroids. The *vasQchip* is connected to a microfluidic pump system, simulating an artificial blood flow and therefore cultivating cells in a more natural environment. Due to the pores in the microchannel exchange between compartments can take place, e.g. to replace or transport nutrients or gases.

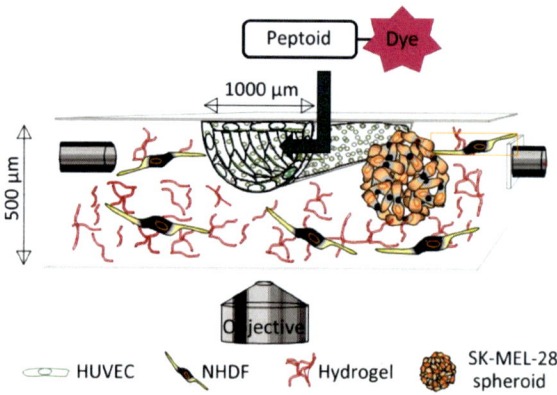

Figure 36: Schematic composition of the *vasQchip* as a cancer model: co-cultivation of HUVEC cells in a porous channel, mimicking a blood vessel, and surrounding tissue, represented by NHDF cells and SK-MEL-28 spheroids embedded in a hydrogel. Fluorescently labeled peptoids are added *via* a microfluidic pump system [168].

This model system gives the possibility to analyze the transport of peptoids through blood vessels to tumor cells as well as their cytotoxic effects not only on spheroids but also on the "healthy" surrounding tissue. Figure 36 shows a schematic composition of the chip system, containing endothelial cells (HUVEC), fibroblasts (NHDF) and spheroids (SK-MEL-28). Uptake of the peptoids in the cells in the *vasQchip* were alanyzed in collaboration with Eva Zittel (Institute of Toxicology and Genetics, KIT). To investigate uptake of peptoids in the different cell lines as a function of time, an exemplary peptoid (P85) was added to the circulation media and cellular uptake was analyzed by confocal fluorescence microscopy after 24 h, 48 h and 96 h.

Brightfield	Hoechst 33342	Peptoid	Merge

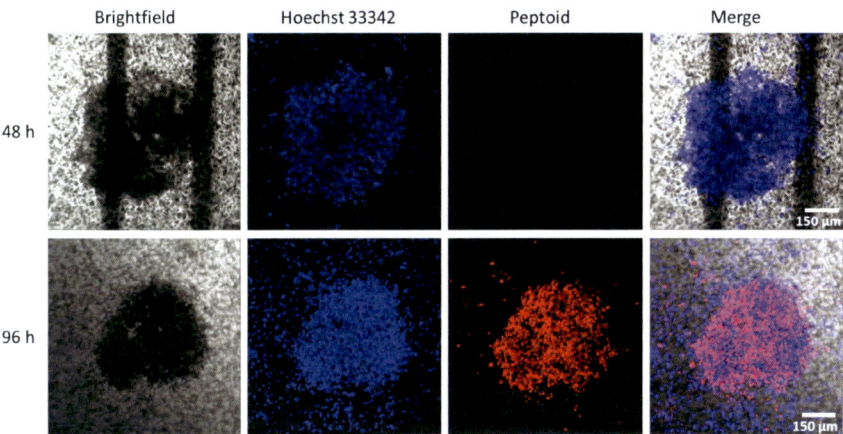

Figure 37: Spheroids (SK-MEL-28, 4 x 10^3 cells) in the *vasQchip*, treated with 10 μM P85 after 48 h and 96 h incubation. Spheroids were analyzed with fluorescence confocal microscopy (Leica TCS-SPE, Objective: ACS APO 10x/0.30 DRY). 1. Brightfield, 2. Hoechst 33342, Ex.: 405 nm, Em.: 410-550 nm 3. Peptoid, Ex.: 561 nm, Em.: 593-696 nm 4. Merge, Scale bar: 150 μm

Strong uptake of peptoids in spheroids was visible after 96 h and fibroblasts, surrounding the spheroid, display low uptake (figure 37). In general, the fibroblasts displayed low peptoid uptake already after 48 h and increased uptake after 96 h, shown in figure 38. As a function of the distance between fibroblasts and the channel, uptake was slightly higher or lower. However, in comparison to spheroids, fluorescence intensity was low. Toxic impacts for fibroblasts could not be determined as cells were confluent, even after 96 h. Increased cytotoxicity was found for HUVEC cells in the channel. Strong uptake was already visible after 48 h and cell density was decreased. After 96 h cell density was very low. Cells, visible in figure 39, are not only HUVEC cells but also fibroblasts directly below the channel, recognizable by their drawn-out shape.

| Brightfield | Hoechst 33342 | Peptoid | Merge |

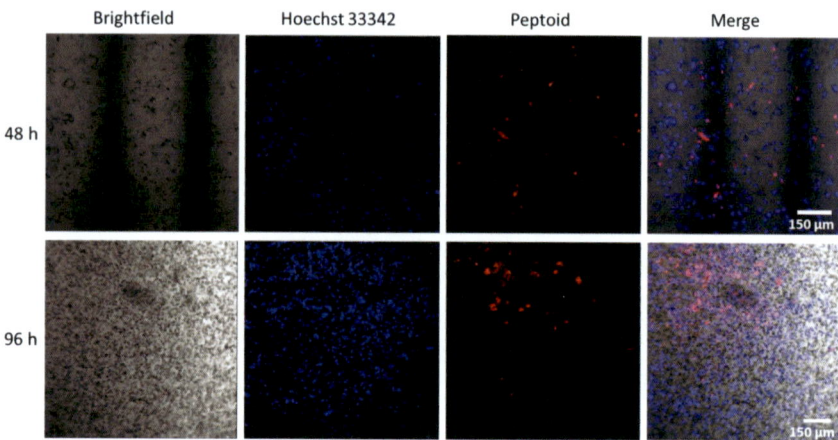

Figure 38: Fibroblasts (NHDF) in the *vasQchip*, treated with 10 µM P85 after 48 h and 96 h incubation. NHDF cells were analyzed with fluorescence confocal microscopy (Leica TCS-SPE, Objective: ACS APO 10x/0.30 DRY) 1. Brightfield, 2. Hoechst 33342, Ex.: 405 nm, Em.: 410-550 nm 3. Peptoid, Ex.: 561 nm, Em.: 593-696 nm 4. Merge, Scale bar: 150 µm

| Brightfield | Hoechst 33342 | Peptoid | Merge |

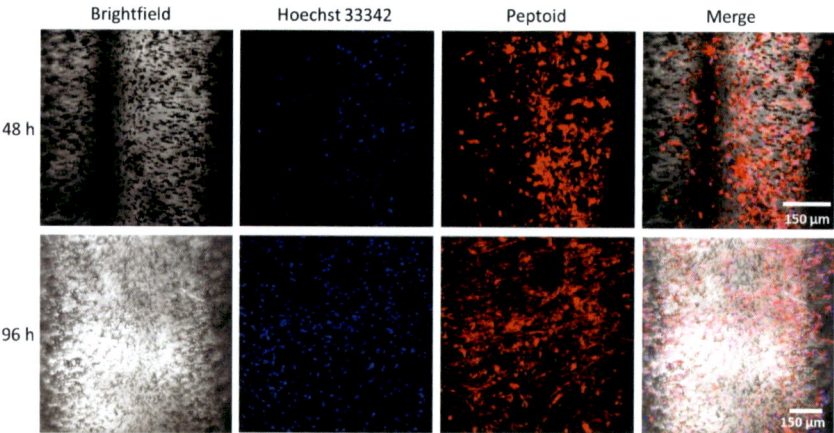

Figure 39: Endothelial cells (HUVEC) in the *vasQchip*, treated with 10 µM P85 after 48 h and 96 h incubation. HUVEC cells were analyzed with fluorescence confocal microscopy (Leica TCS-SPE, Objective: ACS APO 10x/0.30 DRY). 1. Brightfield, 2. Hoechst 33342, Ex.: 405 nm, Em.: 410-550 nm 3. Peptoid, Ex.: 561 nm, Em.: 593-696 nm 4. Merge, Scale bar: 150 µm

These results confirm 2D toxicity assays (chapter 3.2.) that peptoids display higher toxicity to HUVEC cells compared to fibroblasts. However, low cell density of HUVEC cells could also be due to frequent disconnections and connections of the chip from the microfluidic system for confocal microscopy, leading to mechanical stress and pressure to the cells in the channel. Furthermore, rhodamine B, coupled to the tested peptoids, could increase the toxicity to cells. Hence, these experiments have to be repeated and confirmed.

3.2.2. Comparison to peptoids without fluorescent dye

So far, all peptoids which were analyzed as anticancer agents were coupled to rhodamine B for a better visualization in cells and whole organisms e.g. zebrafish embryos. However, rhodamine B itself is assumed to be carcinogenic and even though many anticancer drugs are known to induce cancer, it would be appropriate to use peptoids without a fluorescent dye as a drug. As the influence of dyes on peptoids is still not completely understood, four representative peptoids (P88, P96, P118 and P214), which displayed promising results in experiments described above, were synthesized again without fluorescent labeling. Cytotoxicity and LD_{50} values of P88-A, P96-A, P118-A and P214-A were analyzed in HeLa, HepG2, MCF-7 and NDHF cells (figure 40). Cytotoxicity values were comparable to rhodamine B labeled peptoids, which shows that rhodamine B had no influence on cytotoxicity of cell penetrating peptides towards cancer or primary cell lines.

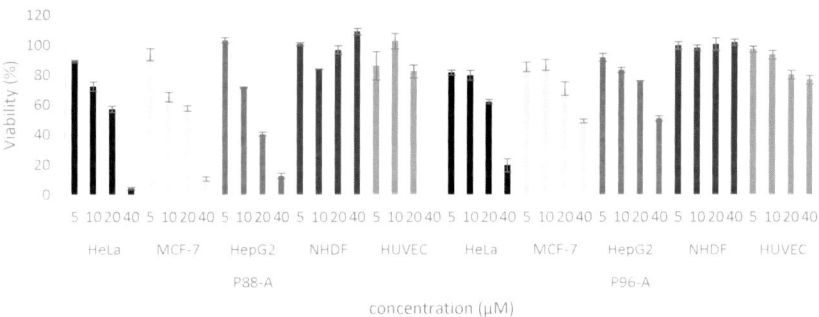

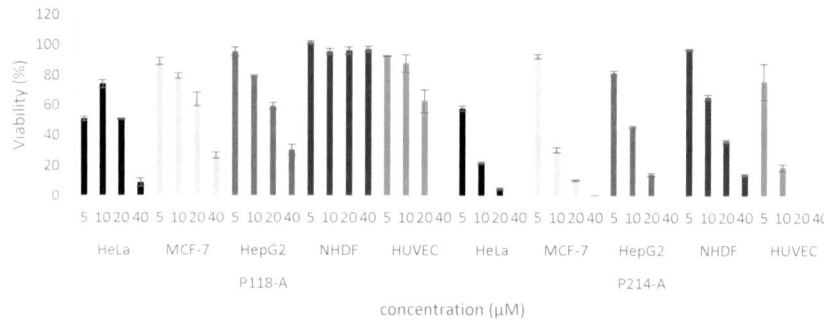

Figure 40: Cytotoxicity of potential anticancer peptoids to primary (NHDF) and cancer (HeLa, HepG2, MCF- 7) cells. P88-A, P96-A, P118-A and P214-A were tested in 5, 10, 20 and 40 µM concentration for 72 h and viability was determined using the MTT assay.

The most toxic peptoid was peptoid P214-A, which was also one of the most cytotoxic rhodamine B labeled peptoids, with a LD_{50} value between 6 and 8 µM for cancer cells. P88-A and P118-A displayed increased cytotoxicity to breast, liver and cervix cells with LD_{50} values between 14 and 26 µM. P96-A displayed the lowest toxicity with LD_{50} values between 25 and 40 µM. Surprisingly, toxicity for primary fibroblast cells was very low. For P88-A, P96-A and P118-A no toxic effects could be measured and LD_{50} values could not be determined ($LD_{50} >$ 40 µM). P214-A was cytotoxic for NHDF cells, however, toxicity was significantly lower compared to cancer cells. These findings confirm the suitability of cell penetrating peptoids as anticancer agents.

Table 9: LD_{50} values (72 h) for P88-A, P96-A, P118-A and P214-A for HeLa, HepG2, MCF-7 and NHDF cells.

Peptoid	LD_{50} HeLa	LD_{50} HepG2	LD_{50} MCF-7	LD_{50} NHDF	LD_{50} HUVEC
P88-A	~ 22.5 µM	~ 17 µM	~ 23 µM	> 40 µM	~ 27.5 µM
P96-A	~ 25 µM	~ 40 µM	~ 40 µM	> 40 µM	> 40 µM
P118-A	~ 20 µM	~ 26 µM	~ 14 µM	> 40 µM	~ 23 µM
P214-A	~ 6 µM	~ 8 µM	~ 8 µM	~ 15 µM	~ 7.5 µM

Cytotoxicity was also investigated as a function of time and therefore viability of cells was measured, for 20 and 40 µM peptoid solution, after 4, 24, 48 and 72 h, shown in figure 41. Fastest killing mechanism was found for P214-A, displaying an exponential drop of cell viability for 20 µM and 40 µM. Furthermore, LD_{50} value could already be determined within the first 4 hours of incubation (~30 µM). For P88-A, P96-A and P118-A toxic effects within the 4 h were low, and P88-A and P96-A showed linearly decreasing viability values. For P118-A a sigmoidal fit was found, with the strongest reduction of viability between 24 h and 48 h.

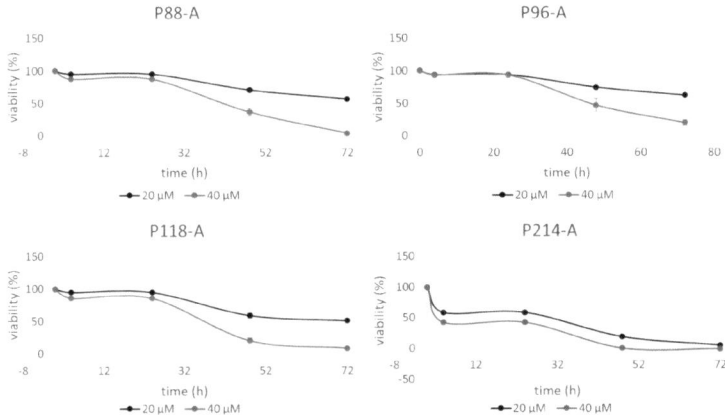

Figure 41 : Viability of HeLa cells after treatment with P88-A, P96-A, P118-A and P214-A (20µM and 40µM) after 4, 24, 48 and 72 h incubation.

In contrast to rhodamine labeled peptoids, pathway of cell death for peptoids without dye could be analyzed by staining of necrotic cells with propidium iodide (PI). In addition, living cells were stained with fluorescein diacetate (FDA). FDA is profluorescent and cell-permeable and is hydrolyzed by intracellular esterases in viable cells to the highly fluorescent fluorescein.

Table 10: Pathway of cell death for 1.5 x 10^4 HeLa and MCF-7 cells incubated peptoids (LD_{50} concentration) for 24 h. Necrotic cells were identified by staining with PI and apoptotic cells by staining with Hoechst 33342 and identification of fragmentation of nuclei. Viable cells were stained with FDA

Peptoid	Apoptotic (HeLa)	Necrotic (HeLa)	Apoptotic (MCF-7)	Necrotic (MCF-7)
P88-A	31%	69%	12%	88%
P96-A	27%	73%	12%	88%
P118-A	33%	67%	23%	77%
P214-A	21%	79%	37%	63%

Unfortunately, for all four peptoids mainly necrotic cell death was found. These findings suit to the investigation of Huang et al., who also determined mainly necrotic pathway of peptoids

[79]. The difference in the pathway compared to rhodamine B labeled peptoids could be due to the increased lipophilicity of four representative peptoids without dye or due to improved analysis by co-staining with PI and FDA. Additionally, intracellular ROS levels were detected by staining cells with H_2DCFDA and the growth measurement of spheroids was repeated for peptoids without rhodamine B. Strong DCF signals were detected for P88-A, P118-A and P214-A. P96-A, which contains, in contrast to P88-A, P118-A and P214-A, only two times Npcb, induced lower levels of ROS (figure 81, appendix). According to these results, it can be assumed, that P88-A, P96-A, P118-A and P214-A also accumulate mainly in mitochondria, as generation of ROS is induced in this cellular compartment. As shown for rhodamine B labeled peptoids, the growth curves of spheroids approve the cytotoxicity results found in the 2D cell culture. Figure 42 displays that P88-A, P118-A and P214-A were highly toxic for cells growing in spheroids, P96-A had very little effects. Increased size of spheroids treated with P88-A and P214-A on day 3 (101% and 88%) can be explained by decreasing cell-cell contacts after peptoid treatment and therefore lose composition of cells in the spheroids.

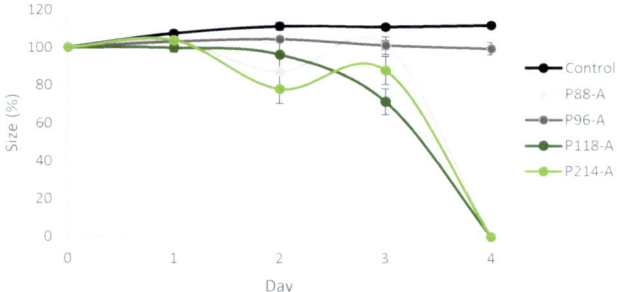

Figure 42: Analysis of growth of SK-MEL 28 spheroids (4 x 10^3 cells) treated with 40 µM potential anticancer peptoids P88-A, P96-A, P118-A and P214-A and non-treated control spheroids. Diameters of spheroids were measured after 1, 2, 3 and 4 days treatment.

Cytotoxicity was also investigated *in vivo* in zebrafish (*Danio rerio*) embryos. The tropical freshwater fish is a frequently used model organism in biological research, e.g. gene analyse and drug screenings [169]. The zebrafish gained intense attractions after the discovery that the zebrafish genome displays around 70% homology to human genes [170]. Homology of organs to humans, in a reduced and simplified form, makes the zebrafish an optimal *in vivo* model for genetics, developmental biology and disease models. To analyse *in vivo* toxicity of peptoid in zebrafish a modification of the fish embryo toxicity test (FET) was used. The normal protocol for FET tests is performed by incubating zebrafish embryos, shortly after fertilization, with various concentrations of substances. Toxicity *in vivo* and impacts on zebrafish development, including hatching rate, deformation rate, movements and development of organs, can be investigated [171]. However, the chorion surrounding embryos before hatching might be a barrier for substances, especially for highly lipophilic peptoids and could therefore lead to false negative results [172]. Thus, zebrafish embryos were dechorionated at 24 hpf stage and afterwards incubated with the peptoid solution. P118-A and P214-A were tested in low micromolar range (5 µM, 10 µM, 25 µM and 50 µM), by incubating 24 hpf dechorionated embryos with peptoid solution in E3-medium (5 mM NaCl, 0.17 mM KCl, 0.33 mM $CaCl_2$, 0.33 mM $MgSO_4$). For each concentration six embryos were tested, by incubating each embryo in a single well containing 100 µl in the respective peptoid concentration. Embryos were incubated at 27 °C and were analyzed at 48 hpf, 72 hpf, 96 hpf and 120 hpf for mortality rates and development by microscope. For P118-A and P214-A no increased toxic effects on zebrafish larvae could be detected. LD_{50} values could not be determined and the highest mortality rate was 33% for P118-A - 5 µM, P214 - 5 µM and P214-A - 50 µM after 24 h incubation. However, 29% mortality was also found for the control, assuming that those embryos might be injured during the dechorionation process, leading to early death.

Table 11: Fish embryo toxicity test of P118-A and P214-A: Dechorionated 24 hpf zebrafish embryos were incubated with peptoids (5 µM, 10 µM, 25 µM and 50 µM) for 4 days and development, mortality and behaviour was observed by microscopic analysis every 24 h.

	Mortality	Deformation	Degradation of movements
P118-A 5 µM	33%	0%	no
P118-A 10 µM	16%	16%	no
P118-A 25 µM	0%	0%	no
P118-A 50 µM	0%	0%	no
P214-A 5 µM	33%	0%	no
P214-A 10 µM	0%	0%	no
P214-A 25 µM	0%	33%	no
P214-A 50 µM	33%	16%	no
Control	29%	0%	no

Figure 43 shows 96 hpf larvae incubated with 50 µM P118-A and P214-A as well as control larvae. For most embryos no differences in development, size and behavior could be observed. For P214-A an increased number of deformed larvae was developing. Those larvae developed a buckled backbone and a slightly malformed head. Malformations seemed to occur randomly and did not increase with higher concentrations. Hence, it cannot be excluded that deformations were accidental and not induced by the peptoid solution.

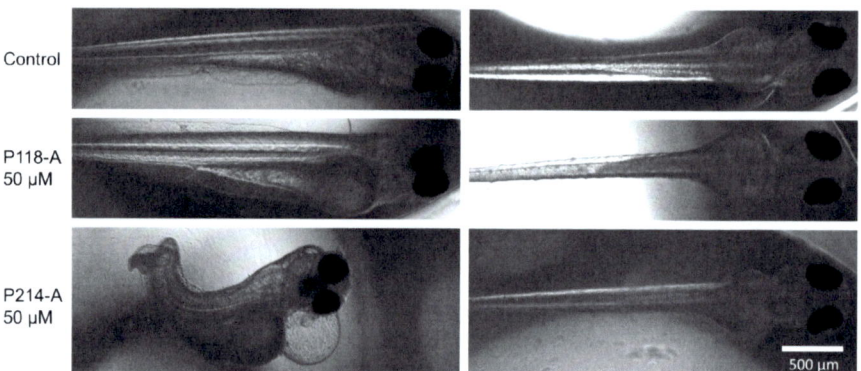

Figure 43: Zebrafish incubated with 50 µM P118-A and P214-A for 4 days and non-treated control zebrafish. Two exemplary zebrafish (lateral and ventral) are shown for each peptoid and control. Images were taken with a Leica DMIL LED microscope (HI PLAN 4x/0.10 DRY Objective), Scale bar: 500 µm

3.3. Screening of peptoid libraries in zebrafish larvae

The zebrafish, *Danio rerio*, a tropical freshwater fish is one of the most important model organisms, not only to analyze gene functions and diseases but also for high-throughput screenings of chemical libraries. Zebrafish are frequently used in large screenings because of their small size, transparency, short reproductive cycle and the availability of many transgenic lines. Their body composition shows a high accordance, with humans, which makes them an optimal organism to screen for molecular transporters. Zebrafish embryos also give the possibility to screen for brain specific transporters as they develop their blood-brain-barrier (BBB) within the first four days, which displays high similarity to the human BBB. F. Rönicke could already show suitability of zebrafish larvae for high-throughput screenings of organ-specific peptoids [91]. Screening of peptoids for organ specificity was performed with *Casper* zebrafish embryos. These fish are doubly mutant for nacre and roy and therefore lack of melanocytes and iridophores [173]. Hence, they lose their typical pigmentation pattern, which makes them optimal for microscopy. Screening was performed with four days old zebrafish larvae, as fish in this age are already hatched and have an almost complete developed organ system, comparable to the human system. For each peptoid eight zebrafish embryos were incubated with a 50 µM peptoid solution in E3-medium for two hours in the dark. Afterwards, fish were washed several times with medium and subsequently anesthetized with 0.02% tricaine and for high-throughput-microscopy zebrafish were transferred in a 96-well IBIDI plate. The screening approach is summarized in figure 44 and can be divided in two major steps. Initially a prescreening, using an Olympus Scan^R IX81 microscope, was performed with a 1.25 x objective to get an overview of the larvae. Afterwards regions of interest (ROI) can be defined by using the MATLAB tool zebrafish Graphical User Interface (zfGUI). This technique was described by Peravali et al. in 2011 and gave the possibility to select several fluorescent regions, by "manual click detection" in the larvae for high resolution images and z-stacks. Coordinates of ROIs were extracted, xml files were generated and in the second step of the process ROIs were imaged again by using a 10 x objective and z-stacks [174]. For all pictures two filters were used, brightfield- and red-filter, to visualize rhodamine B labeled peptoids.

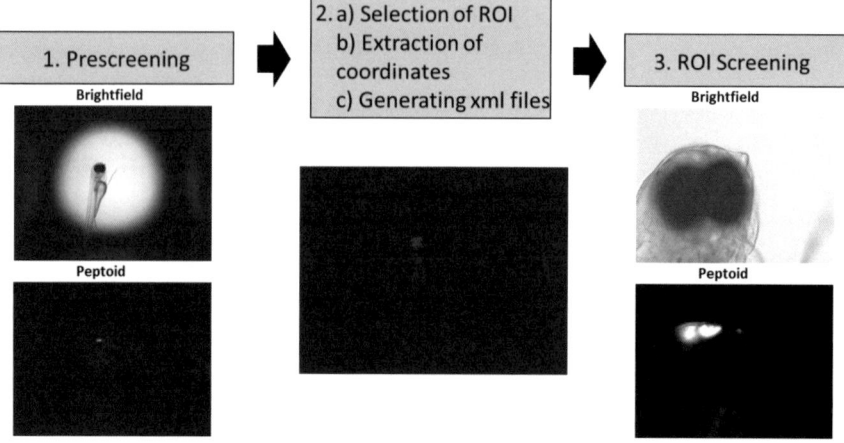

Figure 44: Screening approach for the the investigation of cell penetrating peptoids in zebrafish larvae. 1: Prescreening of all zebrafish in the 96-well IBIDI plate by using an 1.25 x objective. 2. a) Interesting regions (ROI) were defined with the zfGUI tool. 2. b) Coordinates of ROI were extracted and 2.c) xml files were generated. 3) Subsequently, high-resolution images of ROI were done. [91, 174]

Subsequently, zebrafish were analyzed and categorized according to their fluorescent phenotype. Toxicity could be determined by microscopic observation of the heartbeat rate and movements of larvae after treatment with peptoids.

3.3.1. Library 1

For the tetrameric peptoid library 1, containing Nlys, Nprg, Nphe and Npcb, six organ-specific categories were found: digestive system, eye and yolk sac, caudal vein, kidney, lateral line and olfactory system. However, there were also some toxic peptoids leading to apoptotic cells, unspecific accumulation or death of larvae. Most groups were already identified by F. Rönicke for the unpurified peptoids [91]. However, a more defined structure-function relationship analysis could be performed after purification, as false classifications, due to side products could be excluded. In contrast to F. Rönickes findings, toxic peptoids could be subdivided in two groups: peptoids with moderate toxicity, leading to apoptotic cells and highly toxic peptoids, leading to degradation of movements, slower heartbeat or death of zebrafish. In total, 43 peptoids were highly toxic and 26 peptoids displayed moderate toxicity. Uptake of highly toxic peptoids was high in all tissues of the larvae and strong peptoid signals were detected in the blood system. Most occurring side chain, for peptoids in this group, was Npcb with 48% and amount of the hydrophilic side chain Nlys was very low (10%). 42% of peptoids in this group contained Npcb twice and 21% contained it even thrice. There were only two peptoids (5%) in the group which did not contain Npcb. These findings fit very well to the cellular cytotoxicity studies which already displayed increased toxicity of peptoids rich in Npcb. In contrast, peptoids leading to apoptotic cells, represented by unspecific dot pattern all over the zebrafish body, were rich in hydrophilic and hydrophobic side chains [175]. Occurrence of Nlys and Npcb was almost identical with 34% and 36%. Four, out of six peptoids, containing two times Nlys and two times Npcb were present in this group. Exemplary microscopy images of larvae, distribution of side chains and sequences of ten representative peptoids in the group are shown in figure 45.

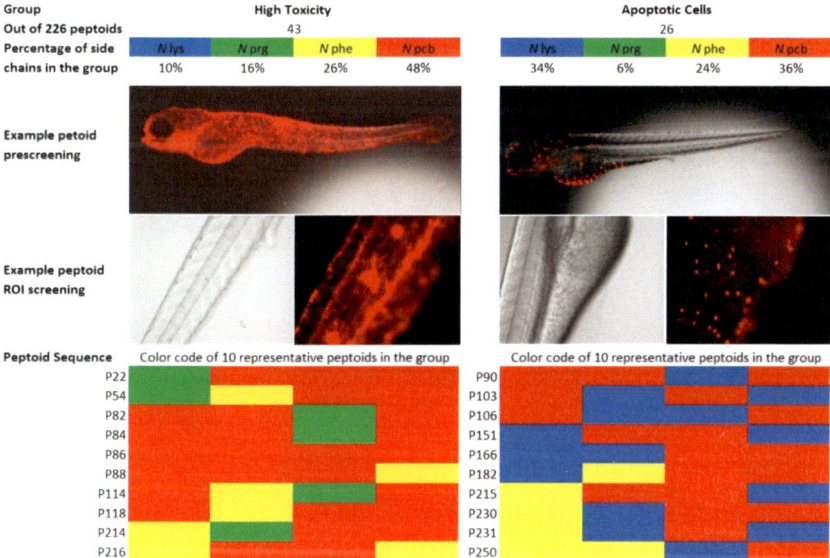

Figure 45: Overview of the subgroups "high toxicity" and "apoptotic cells" according to the classification of the fluorescent phenotype of the peptoid library after automated screening of zebrafish larvae. Amount of peptoids in the subgroup, ratio of side chains, exemplary images of the phenotype (prescreening and ROI screening) and ten selected sequences for the respective subgroup in color code are shown.

The group of peptoids accumulating in the eye (lens) and yolk sac was rather small, containing only 13 peptoids. Due to the high homology of the zebrafish eye to the human eye and their large size in comparison to the overall size of zebrafish larvae, the zebrafish is an important model to understand human eye diseases [176-178]. Zebrafish embryos develop their eyes within the first days and at 72 hpf stage visual response and an almost completely developed retina can be found [179]. Peptoids accumulating in zebrafish eyes were rich in *N*phe (33%), while the hydrophilic monomer *N*lys was the lowest occurring side chain (15%). All four variations of three times *N*prg and one-time *N*phe were present in this group. Hence, it can be assumed, that only the ratio of the respective side chain in a peptoid had an effect on the localization. The exact position of the side chain was not relevant. The category digestive system represents 27 peptoids. Intestine anatomy of zebrafish is also comparable to the mammalian small intestine [180, 181]. Therefore, zebrafish are used to study diseases like diabetes and colorectal cancer. The zebrafish mouth is already open 72 hpf and most parts of the digestive system are developed within the first 96 h, however, first feeding and opening of the gastrointestinal tube starts on day 6 [180]. Peptoids in the digestive system were rich in *N*prg, which was appearing with 43%. Uptake of those peptoids was low and probably took place only by swallowing of the peptoid solution. An overview of both categories, eye and yolk sac and digestive system, is presented in figure 46.

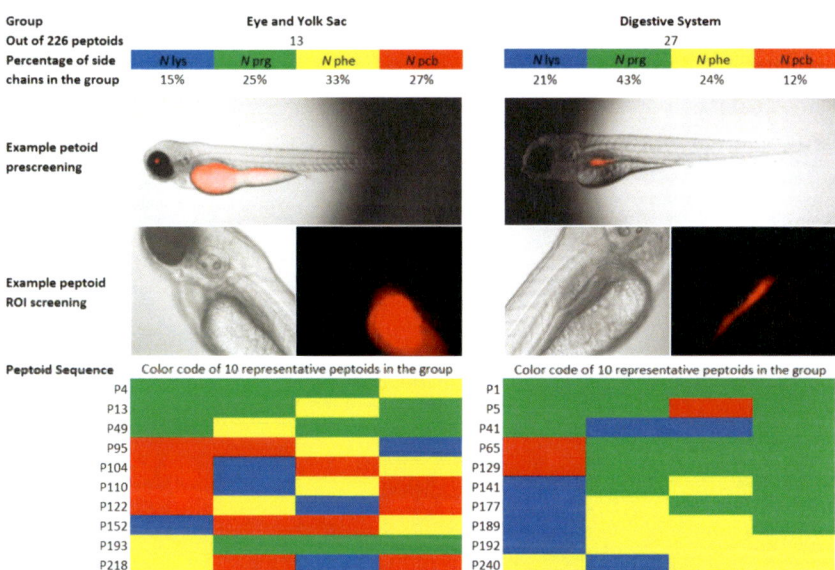

Figure 46: Overview of the subgroup "eye" and "yolk sac" and "digestive system" according to the classification of the fluorescent phenotype of the peptoid library after automated screening of zebrafish larvae. Amount of peptoids in the subgroup, ratio of side chains, exemplary images of the phenotype (prescreening and ROI screening) and ten selected sequences for the respective subgroup in color code are shown.

A rather hydrophilic group of peptoids was found in the caudal vein. Blood vessels have essential functions for zebrafish, as well as for vertebrates, as oxygenated blood is transported throughout the body. Additionally, it serves as communication system between different organs, by transport of hormones [182]. Peptoids accumulating in the caudal vein also displayed a slightly cytotoxic impact on zebrafish larvae. This was visible due to the strong accumulation of the peptoids also in other parts of the body (lateral line, gills, and apoptotic cells). In addition, they expressed impacts on the movements of larvae. Regarding the composition of peptoids in this group, they were comparable to peptoids which were found to induce apoptotic cells. However, peptoids inducing apoptotic cells had higher ratios of aromatic side chains to cationic side chains. A small category was displayed by 7 peptoids accumulating in the kidney. The pronephric kidney of zebrafish larvae is less complex compared to the human kidney. In embryonic stage, it is composed of two parallel nephrons with glomeruli fused at the midline [183, 184]. Peptoids were probably only accumulating in the head kidney, at the anterior part of the kidney, composed of pronephric glomerulus and interrenal tissue [185]. Four peptoids in this category were composed of all four side chains, without doubling one side chain. However, no clear structure-function relationship could be observed as the three remaining peptoids in this subgroup were very different in sequence. An overview of both subgroups is shown in figure 47.

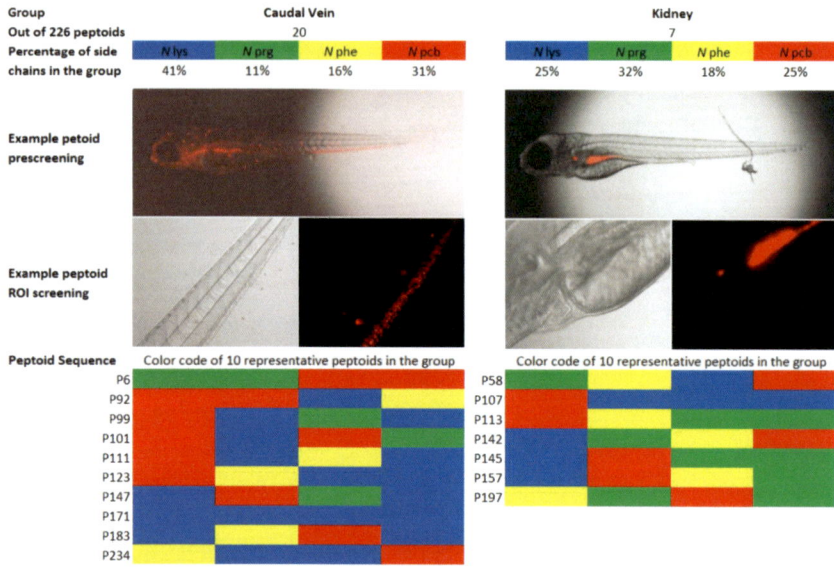

Figure 47: Overview of the subgroup "caudal vein" and "kidney" according to the classification of the fluorescent phenotype of the peptoid library after automated screening of zebrafish larvae. Amount of peptoids in the subgroup, ratio of side chains, exemplary images of the phenotype (prescreening and ROI screening) and ten selected sequences for the respective subgroup in color code are shown.

Two very different groups were peptoids accumulating specifically in the two sensing organs, the olfactory system and the lateral line, which are both rich in neurons, shown in figure 48. While peptoids accumulating in the lateral line were very hydrophilic, with 39% Nlys and only 12% Npcb, peptoids in the olfactory system were hydrophobic and Nlys was not present in this subgroup. Aquatic vertebrates use the lateral line to detect movements, differences of pressure or vibrations to recognize fellows, prey animals or enemies. The major component of the lateral line are small sensory patches, the neuromasts, consisting of hair cells, which are modified epithelial cells. These patches are distributed all over the body surface and position and amount depends on the species. In 96 hpf zebrafish eight neuromasts can be found [186]. Peptoids accumulating in the lateral line displayed the group with the highest number of peptoids found for library 1. 29% of the peptoids in library 1 were found in this group, with a wide variety of side chain compositions. Nlys was found in 98% of the peptoids, in this subgroup, at least once. Hence, the cationic character of the peptoid was probably triggering the uptake in neuromasts. 40% of the peptoids contained Nlys twice and 11% contain it thrice. The amount of aromatic and lipophilic side chains was low with 21% for Nphe and 12% for Npcb.

Peptoids accumulating in the olfactory system were especially rich in Nprg and Nphe, with 37% and 38%. The highly lipophilic side chain Npcb, was less present with 26% and only 20% contained more than one-time Npcb, as peptoids rich in Npcb were mostly found to have toxic impacts. Nprg and Nphe seem to play an essential role for accumulation in the olfactory system, as most peptoids contained Nprg and Nphe at least once (92%). 56% of the peptoids in this group contained Nphe twice and 44% contained Nprg twice. The olfactory system is an important sensory organ for zebrafish to sense food, enemies or potential mates. In contrast to humans, zebrafish are able to smell amino acids, gonadal steroids, prostaglandins and bile acids, with their olfactory sense [187]. A paired olfactory organ is located between eyes and forebrain, consisting of a cavity with two openings, acting as entrance and exit of the water flow. The olfactory rosette is located between those nares, consisting of sensory and non-sensory epithelia. The olfactory receptor neurons, ciliated olfactory or microvillar olfactory neurons, are located within the sensory epithelia. Axons of the olfactory receptor neurons are united to the olfactory nerve, which is leading into the olfactory bulb [188]. A schematic and simplified representation of the olfactory system in zebrafish embryos is shown in figure 50. The olfactory nerve is already present in 24 hpf zebrafish embryos and by 48 hpf it is enlarged and the optic nerve and microvillous neurons are added [189].

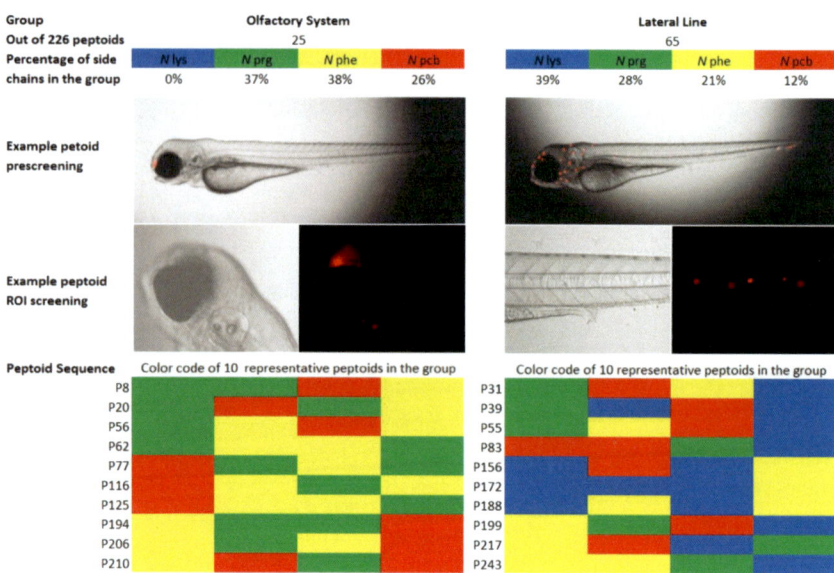

Figure 48: Overview of the subgroup "olfactory system" and "lateral line" according to the classification of the fluorescent phenotype of the peptoid library after automated screening of zebrafish larvae. Amount of peptoids in the subgroup, ratio of side chains, exemplary images of the phenotype (prescreening and ROI screening) and ten selected sequences for the respective subgroup in color code are shown.

Furthermore, the impact of the exact position of the particular side chains in the peptoids, which localized to the olfactory system and the lateral line, was analyzed as shown in figure 49. For peptoids accumulating in the olfactory system the ratio between aliphatic side chains (Nprg) and aromatic side chains (Nphe and Npcb) was equal in position 1, 2 and 4 (1:2.1). For position 3 the ratio between aromatic and aliphatic side chains was almost equal (1:1.3). Npcb, was occurring more frequently in outer positions 1 and 4, compared to the position 2 and 3. For peptoids accumulating in neuromast cells, an increasing amount of Nlys from position 1 to position 4 was found. While 29% of the peptoids in this subgroup contained Nlys in position 1, 43% contained Nlys at position 4. In contrast to that, the amount of Nphe was decreasing from position 1 to position 4. The amount of peptoids, containing Nphe at position 1 was high, with 45%, but very low at position 4, with 17%. The ratio of Nprg and Npcb was nearly consistent with approximately 28% and 12%.

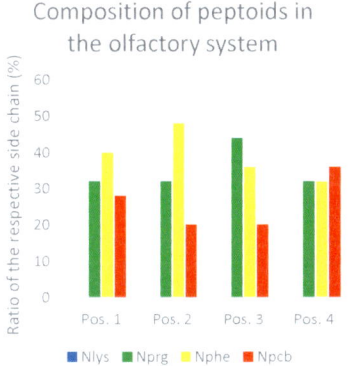
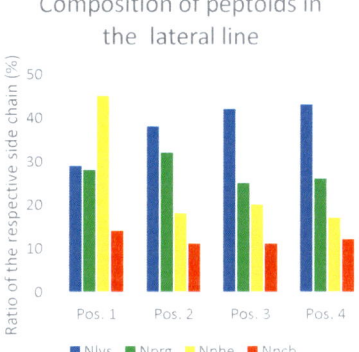

Figure 49: Composition of peptoids in olfactory system (left) and lateral line (right): Ratio of side chains (Nlys, Nprg, Nphe and Npcb) on position 1, 2, 3 and 4 in the tetrameric peptoid is shown.

Accumulation of peptoids in the olfactory system is especially interesting in terms of drug delivery to the brain. Delivery of drugs into the brain is challenging due to the BBB, which is a major obstacle for hydrophilic and large molecules. Tight junctions close inter-endothelial pores and therefore prevent free diffusion of molecules across the barrier. The olfactory region provides a potential pathway for the transport of drugs in the CNS avoiding the BBB [190-192]. For example for viruses, such as poliomyelitis or vesicular stomatitis, the nasal pathway into the brain was firstly discovered [193-196]. Later, it was also found that metals, like gold, cadmium and aluminum, are able to enter the CNS *via* the olfactory system [197-199]. Sakane *et al.* analyzed cerebrospinal fluid (CFS) transport of drugs *via* the nasal cavity as a function of their lipophilicity and could show increased CFS delivery with increased lipophilicity of drugs. Peptoids accumulating in the olfactory system are highly lipophilic with theoretical LogP values with an average of 3.5 (e.g. P8: 1.95, P94: 5.8, P116: 3.19, P210: 4.63, calculated with *Molinspiration* [119]), assuming that they are efficient transporters for drug delivery to the brain *via* the nasal cavity.

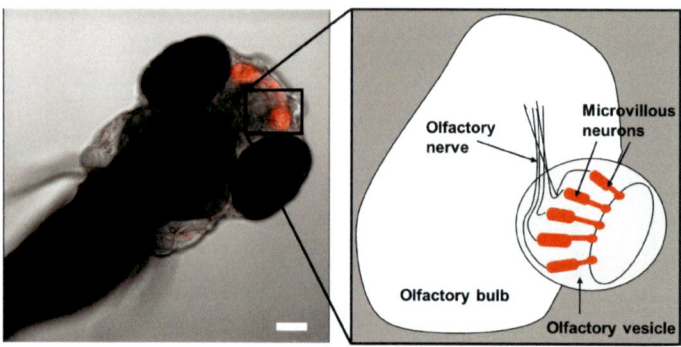

Figure 50: Schematic view of the olfactory system in zebrafish. Left: 96 hpf zebrafish incubated with P116, and analyzed with fluorescence confocal microscopy (Leica TCS-SPE, Objective: ACS APO 10x/0.30 DRY, Scale bar: 100 µm). Right: Composition of the olfactory system, Source: [200]

According to the literature, three pathways for nasal administered substance are possible: the olfactory nerve pathway, the olfactory epithelial pathway and the systemic pathway (figure 51). The systemic pathway is an indirect pathway into the brain and occurs for substances entering the blood capillaries in the submucosa tissue. However, to reach the CFS or brain tissue crossing the BBB is necessary, hence this pathway is exclusively taken by small, lipophilic molecules. The olfactory epithelial pathway was found for substances entering the olfactory vesicle at some point, other than the neurons, e.g. through supporting cells or diffusion between cell junctions. After crossing the basal membrane, substances can travel *via* the perineural space, surrounding the olfactory nerve, to the CFS and finally into the brain. Last but not least, substances can accumulate in the olfactory receptor neurons and subsequently reach the brain *via* the olfactory nerve [190, 197, 201, 202].

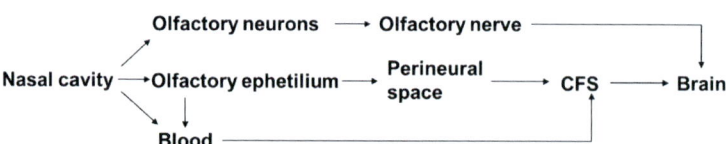

Figure 51: Three possible delivery routes from the nasal cavity to the brain. Source: [201, 202]

Different uptake pathways have been intensively studied for various substances, including viruses, metals, antibiotics and hormones, suggesting that one substance may follow more than one route. However, for most drugs exact routes are still unclear and may involve more steps as known, so far. It can be assumed that peptoids enter the brain *via* the olfactory nerve pathway, as it has been shown that they accumulate in olfactory receptor neurons as well as in low concentrations in the olfactory nerve [91].

3.3.1.1. Injection of peptoids into cardinal vein

For the high-throughput investigation of peptoid libraries in zebrafish larvae, to identify organ-specific transporters, application of peptoids *via* bathing is a suitable method. Large peptoid libraries can be evaluated in embryos and structure-function relationships can be done. After evaluation of all peptoids, toxic peptoids or peptoids without organ specificity can be excluded and further investigation can be done with interesting candidates. Screening of peptoid library 1 revealed six interesting categories: eye, digestive system, lateral line, caudal vein, kidney and olfactory system. Most interesting for drug delivery are peptoids accumulating in the olfactory neurons as they would give rise to a potential brain delivery due to the connection of olfactory system and brain. After bathing, some peptoids could also be detected in the olfactory nerve, however, there was no further detection in the brain possibly due to concentration gradients in the brain [91]. A possibility to increase the peptoid concentration in the blood stream of zebrafish, without increasing the toxicity, is microinjection of peptoid solution in the cardinal vein. Two peptoids, shown in figure 52, P116 (*N*pcb-*N*phe-*N*prg-*N*phe-RhodB) and P171 (*N*lys-*N*lys-*N*lys-*N*lys-RhodB), were analyzed for organ specificity after microinjection.

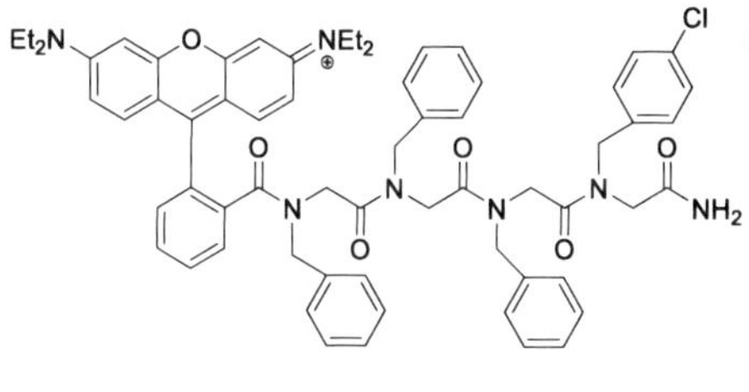

P116 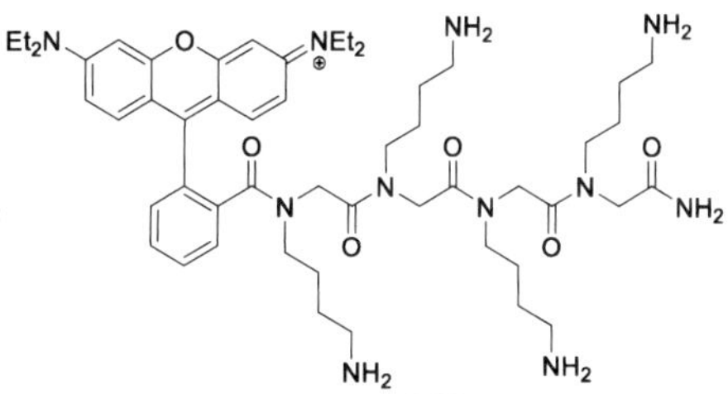   P171

Figure 52: Two peptoids which were injected into the zebrafish blood circulation: Olfactory system specific peptoid P116 (*N*pcb-*N*phe-*N*prg-*N*phe-RhodB) and hydrophilic control peptoid P171 (*N*lys-*N*lys-*N*lys-*N*lys-RhodB).

Brightfield	Peptoid	Merge

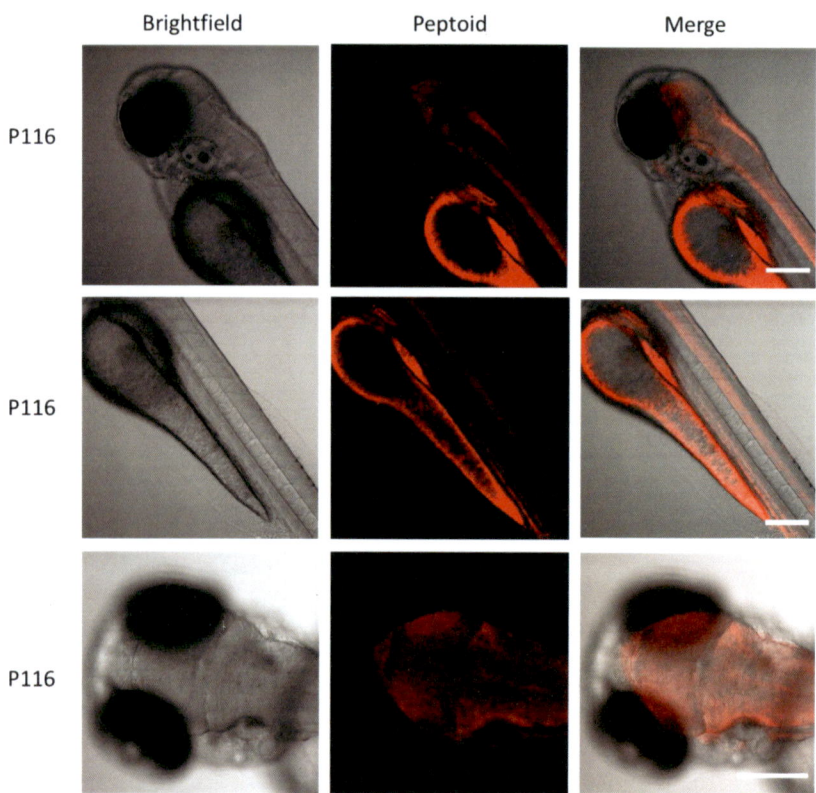

Figure 53: Ventral and lateral view of 3 dpf zebrafish, 24 h after injection of P116 (*N*pcb-*N*phe-*N*prg-*N*phe-RhodB) into the caudal vein, Leica TCS-SPE, objective: ACS APO 10x/0.30 DRY. 1. Peptoid, Ex.: 560 nm, Em.: 590-690 nm 2. Brightfield 3. Merge, Scale bar: 200 μm

While P116 is a lipophilic, representative peptoid for olfactory accumulation, P171 is very hydrophilic and accumulated mainly in caudal vein after bathing of embryos. 48 hpf zebrafish larvae were dechorionated and anesthetized with tricaine. Subsequently, larvae were injected with 1 mM peptoid in E3-medium and afterwards transferred again in E3-medium for 24 h for recovery. Following, larvae were anesthetized again and peptoid localization was analyzed by confocal microscopy. For ventral view of the brain zebrafish were embedded in 1.5% low-melting agarose. Promising results were found for lipophilic peptoid P116, displaying accumulation in brain tissue after injection. Furthermore, accumulation was found in spinal cord and blood vessels (figure 53).

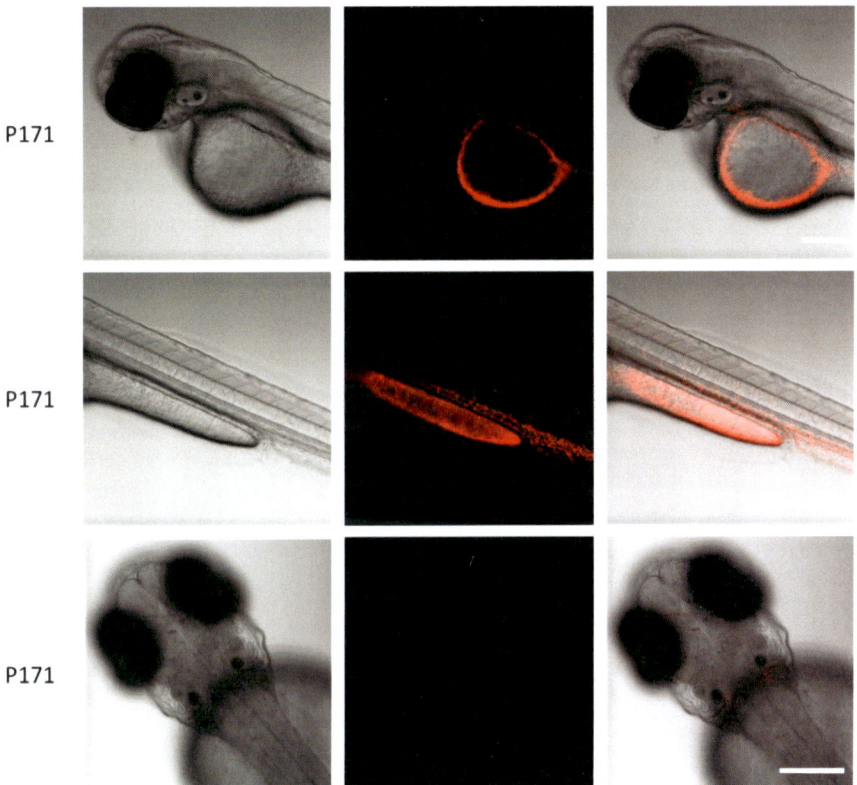

Figure 54: Ventral and lateral view of 3 dpf zebrafish, 24 h after injection of P171 (*N*lys-*N*lys-*N*lys-*N*lys-RhodB) into the caudal vein, Leica TCS-SPE, objective: ACS APO 10x/0.30 DRY. 1. Peptoid, Ex.: 560 nm, Em.: 590-690 nm 2. Brightfield 3. Merge, Scale bar: 200 µm

In contrast, the hydrophilic peptoid P171 was mainly accumulating in endothelial cells of the blood vessels, depicted in figure 54, and brain localization could not be detected. For the analysis of P116 accumulation on cellular level, magnified images of brains were taken. Figure 55 displays the uptake of P116 in brain cells, probably accumulating in cytosol, as well as in brain ventricle and blood vessels of the brain tissue.

P116   Brightfield   Merge

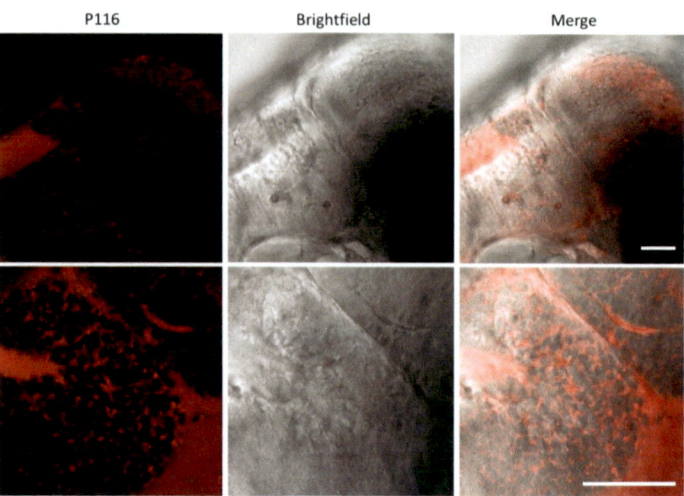

Figure 55: Magnified view of brain of 3 dpf zebrafish, 24 h after injection of P116 (*N*pcb-*N*phe-*N*prg-*N*phe-RhodB) into the caudal vein, Leica TCS-SPE, objective: ACS APO 10x/0.30 DRY. 1. Peptoid, Ex.: 560 nm, Em.: 590-690 nm 2. Brightfield 3. Merge, Scale bar: 50 µm

These findings confirm the suitability of lipophilic peptoids, accumulating in the olfactory system for brain-specific drug delivery. However, the BBB is not fully developed at 72 hpf embryonic stage, therefore, analysis in adult zebrafish have to be done.

3.3.2. Library 2

High-throughput screening of library 2 in zebrafish larvae revealed only three categories. The number of categories for this library was lower due to the fact that library 2 contained less compounds and was less diverse as only three aromatic side chains were permutated. The biggest subgroup were peptoids accumulating in the olfactory system, containing almost 50% of the peptoids in library 2 (figure 56). For library 1 it was shown that *N*phe was highly important for a specific accumulation in the olfactory system. However, in library 2 the *N*phe content was rather small for this group with only 20%. 30% of peptoids do not even contain *N*phe. In contrast, the amount of the polar aromatic side chain *N*pob was high and frequently present with 48%. As shown for library 1, peptoids in the olfactory system were highly lipophilic and displayed an averaged LogP value of 3.5 (theoretical, [119]). The olfactory peptoids in library 2 showed a similar LogP value of 3.8 (theoretical, [119]).

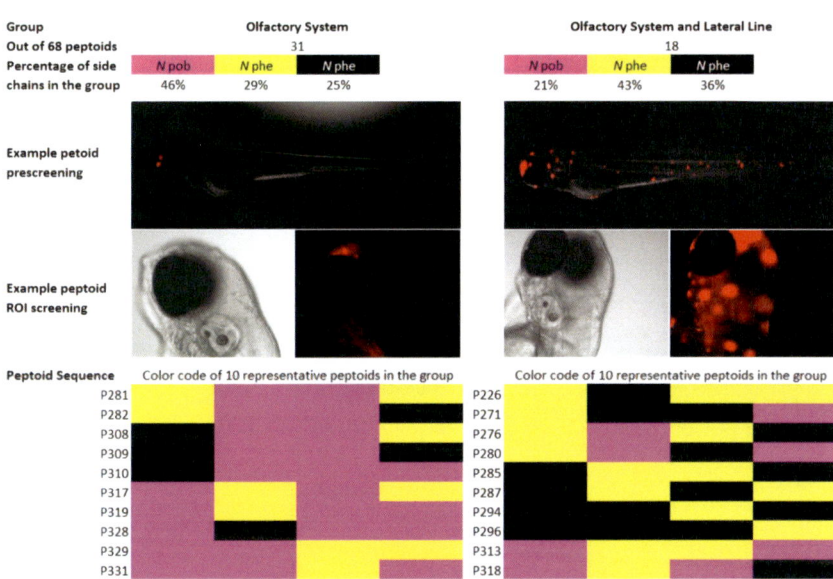

Figure 56: Overview of the subgroup "olfactory system" and "olfactory system and lateral line" according to the classification of the fluorescent phenotype of the peptoid library after automated screening of zebrafish larvae. Amount of peptoids in the subgroup, ratio of side chains, exemplary images of the phenotype (prescreening and ROI screening) and ten selected sequences for the respective subgroup in color code are shown.

Interestingly, for library 2, a new fluorescent phenotype was detected: peptoids which were not only accumulating in the olfactory system but simultaneously in the neuromast cells. For library 1 it was shown, that lipophilic peptoids, accumulating in the olfactory system were completely different in their composition to the hydrophilic peptoids localizing to the neuromast cells of the lateral line. Peptoids in this subgroup were more lipophilic compared to peptoids found only in the olfactory system and the most frequent side chain in this group was *N*phe with 43%. Each peptoid in this subgroup contained *N*phe at least once and even 56% of the peptoids contained it at least twice. *N*pbf was also present in most peptoids, with 89% containing it at least once and 39% containing it at least twice. The fewest occurring side chain in this subgroup was *N*pob with 21%, present in 61% of the peptoids. Hence, the affinity of peptoids to neuromast cells was not preserved exclusively to hydrophilic peptoids, containing positive charged side chains. Highly lipophilic peptoids, rich in *N*phe and *N*pbf, were also found to accumulate in the neuromast cells.

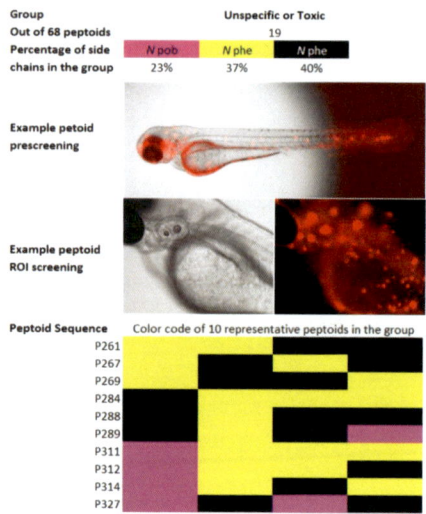

Figure 57: Overview of the subgroup "unspecific" or "toxic" according to the classification of the fluorescent phenotype of the peptoid library after automated screening of zebrafish larvae. Amount of peptoids in the subgroup, ratio of side chains, exemplary images of the phenotype (prescreening and ROI screening) and ten selected sequences for the respective subgroup in color code are shown.

28% of the peptoids in library 2 were found to be toxic for zebrafish larvae, leading to inhibition of movements, low hearth beat and unspecific accumulation of the peptoid. Those peptoids were leading to apoptotic cells, accumulation in neuromast cells, olfactory system, blood vessels, digestive system and gills, shown in figure 57. Toxicity was probably caused by the lipophilic side chain Npbf, which was present with 40% in this subgroup. 89% of the peptoids in this subgroup contained Npbf at least once, 58% contained it at least twice. As the lipophilic side chain Npbf had similar toxic effects as Npcb in library 1 in can be concluded, that highly lipophilic side chains are only suitable for transporter peptoids in a moderate amount and should not exceed a ratio of 1:1 compared to hydrophilic side chains or those with moderate lipophilicity. Npob was the fewest occurring side chain, with 23%, in this subgroup, present in 76% of the peptoids. Most peptoids contained it only once (52%) and only few peptoids contained it twice (21%).

3.3.3. Library 3

Library 3 consisted of decameric and dodecameric peptoids as presented in chapter 3.3.. Unfortunately, organ-specific accumulation could not be detected for those peptoids. Even though they are rich in Nlys, usually triggering the uptake in neuromast cells for tetrameric peptoids, only unspecific accumulation in the larvae were found. Decameric peptoids P322, P323, P324 and dodecameric P327 were strongly toxic, leading to the death of larvae after 2 h exposure to peptoid solution. Confocal z-stack images for the remaining zebrafish, incubated with P323, P326, P328 and P329 are shown in figure 58. P325 and P329 displayed similar uptake, mainly in skin cells. Structure of P325 (Nphe-Nphe-Nbut-Nlys-Nlys-Nlys-Nlys-Nbut-Nphe-Nphe-RhodB) and P329 (Nphe-Nphe-Nbut-Nlys-Nlys-Nlys-Nlys-Nbut-Nphe-Nphe-Npcb-Npcb-RhodB) is related to each other, including Nphe, Nbut and Nlys, with the only variation that P329 contained to additional Npcb. P328 (Nphe-Nphe-Nhep-Nlys-Nlys-Nlys-Nlys-Nhep-Nphe-Nphe-Npcb-Npcb-RhodB) showed high uptake in the whole organism and P326 (Nlys-Nphe-Nphe-Nlys-Nphe-Nphe-Nlys-Nphe-Nphe-Nlys-Npcb-Npcb-RhodB) was leading to apoptotic cells. Apoptotic cells, caused by P326, were probably due to the mixture of strongly hydrophobic and hydrophilic side chains, as it was shown in chapter 3.3.1. for tetrameric peptoids rich in Nlys, Nphe and Npbc.

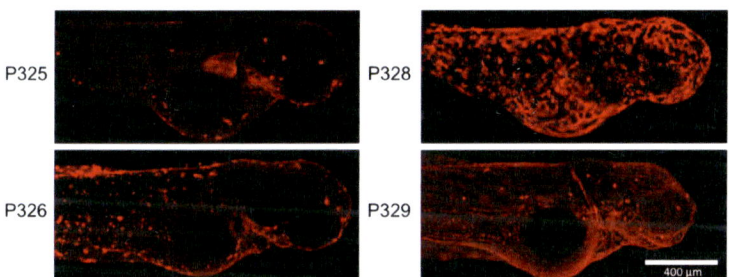

Figure 58: 4 dpf zebrafish, incubated for 2 h with P325, P326, P328 and P329, Leica TCS-SPE, Objective: ACS APO 10x/0.30 DRY. 1. Peptoid, Ex.: 560 nm, Em.: 590-690 nm 2. Brightfield 3. Merge, Scale bar: 400 µm

3.3.4. Library 4

Cyclic peptoids and their linear counterparts were also investigated for organ specificity and differences in uptake *in vivo*. As described for library 1 and library 2, zebrafish were incubated for 2 h with peptoid solution and subsequently screened with an Olympus Scan^R IX81 microscope. Zebrafish incubated with cyclic peptoids and their linear counterparts are shown in figure 59. Cyclo 1 (*N*phe-*N*prg-*N*phe-*N*phe-*N*4az-*N*phe) accumulated in neuromast cells and olfactory system, in contrast, linear 1 displayed increased toxicity and no specific uptake. Uptake of cyclo 2 and linear 2 (*N*phe-*N*prg-*N*pob-*N*pob-*N*4az-*N*phe) was low in both cases. However, cyclo 2 accumulated in the olaftory system, while linear 2 showed unspecific uptake leading to apoptotic cells. Uptake in the olfactory system of cyclo 2 also fits to the results found for the embryo screening of library 2. In this analysis it was shown that peptoids rich in *N*pob expressed specific uptake in the olfactory system. Unexpectedly, the hydrophilic cyclo 3 (*N*phe-*N*prg-*N*lys-*N*lys-*N*4az-*N*phe) accumulated in the olfactory system as well as in the caudal vein. Furthermore, unspecific uptake was found for cyclo 3. In contrast, linear 3 showed exclusively unspecific uptake. Big differences were found for cyclo 4 and linear 4 (*N*phe-*N*prg-*N*pob-*N*pob-*N*4az-*N*pob). Linear 4 was accumulating unspecifically in neuromast cells, gills, veins and apoptotic cells, while accumulation of cyclo 4 was found specific in olfactory system and yolk sac, probably due to its high content of *N*pob. Peptoid 6 (*N*phe-*N*prg-*N*phe-*N*phe-*N*4az-*N*pfb) and Pepoid 7 (*N*phe-*N*prg-*N*pfb-*N*lys-*N*4az-*N*pfb) displayed a toxic impact on larvae for linear peptoids as well as for their cyclic counterparts. However, toxicity was higher for linear peptoids, leading to death of larvae incubated with linear 6 and strong unspecific uptake for linear 7. Cyclo 6 and cyclo 7 and showed similar uptake, even though cyclo 7 was more hydrophilic, in olfactory system, veins, neuromast cells, gills and apoptotic cells. Increased toxicity of peptoid 6 and peptoid 7 might be triggered by high content of *N*pbf and *N*phe, which induced high unspecific uptake, as shown for peptoid library 2 in chapter 3.3.2.

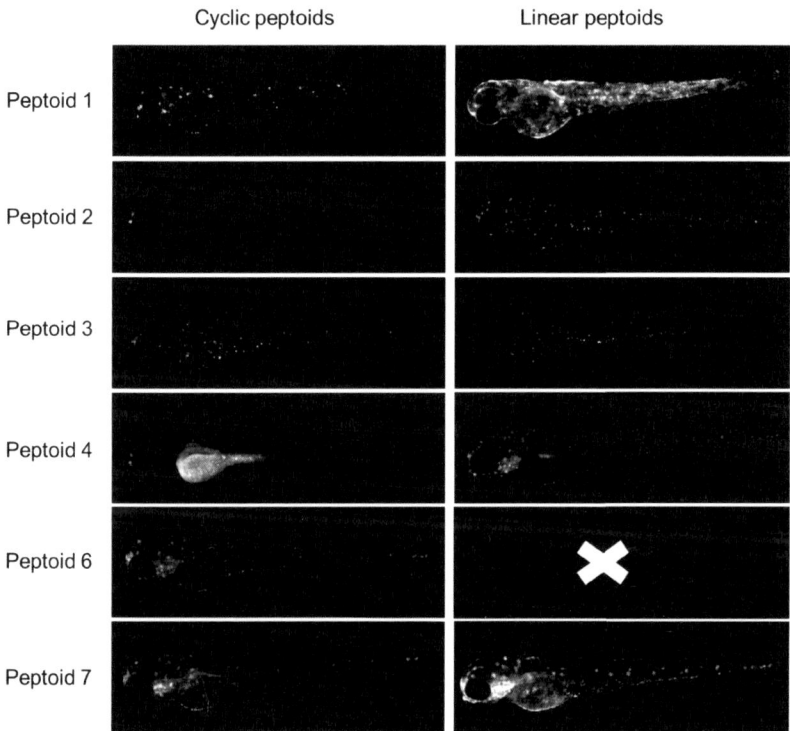

Figure 59: 4 dpf zebrafish, incubated for 2 h with cyclo 1, linear 1, cyclo 2, linear 2, cyclo 3, linear 3, cyclo 4, linear 4, cyclo 5, linear 5, cyclo 6, linear 6, cyclo 7 and linear 7, Olympus Scan^R IX81 microscope.

In summary, the comparison between cyclic and linear peptoids shows lower toxicity and less unspecific uptake of cyclic peptoids. Furthermore, organ specificity of hexameric, linear and cyclic, peptoids matched to the results found for the different side chains in tetrameric peptoids, with an exception of cyclo 3, which was found in the olfactory system although it contained Nlys.

3.3.5. Screening of peptoids in zebrafish embryos using droplet-microarray platforms

Even though many steps in the application and screening process of CPPo analysis in zebrafish embryos is already done *via* high-throughput approaches, treatment of embryos with peptoid solution was still performed manually and was quite time consuming. Droplet-microarrays (DMA), based on patterns of hydrophilic spots separated by superhydrophobic regions, are a new method for high-throughput investigations of substance libraries in zebrafish embryos [203]. Aqueous solutions spontaneously form separated droplets on DMA platforms, due to high affinity of water to hydrophilic spots and strong water repellency of hydrophobic barriers (discontinuous dewetting). Embryos can be spread on platforms, getting only one embryo per spot, in small volumes of 5 µl/spot. Schematic spreading of embryos and embryos on platform are shown in figure 60.

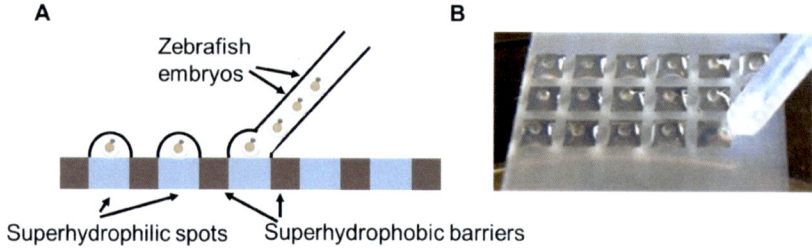

Figure 60: A: Schematic representation of the spreading of zebrafish embryos on DMA plates containing superhydrophilic spots and superhydrophobic barriers B: Image of the spreading process [203].

Afterwards this DMA platform could be used to incubate embryos with different compounds and different compound concentrations. In collaboration with Anna Popova (Institute of Toxicology and Genetics, KIT) eight representative peptoids were screened for organ-specificity on DMA plates and compared to zebrafish incubated in 96-well plates to demonstrate a proof-of-principle. Therefore, peptoids were printed on a microscope glass slide, corresponding to the spots on the DMA plates hosting the fish, in different concentrations. After drying of the peptoid plate and spreading of 24 hpf zebrafish embryos on DMA platform peptoids could be added to droplets by a previously reported sandwiching method [204].

Subsequently, CPPos were added to the droplets, containing the zebrafish embryos, by sandwiching the DMA slide with the glass slide for 15 min. After 24 h of incubation in droplets embryos were collected in 96-well plates, as hatching in droplets is not possible due to the size of the droplets. After further 24 h incubation, allowing the embryos to hatch, zebrafish were screened using an Olympus Scan^R IX81 fluorescence microscope. Peptoid accumulation of zebrafish incubated on DMA plates was matching to experiments on the 96-well plate as shown in chapter 3.3.1., confirming that DMA platforms and treatment *via* sandwiching method are a suitable approach for screenings of CPPo libraries. Figure 61 shows accumulation of P208 in the olfactory system for zebrafish incubated with the two different methods.

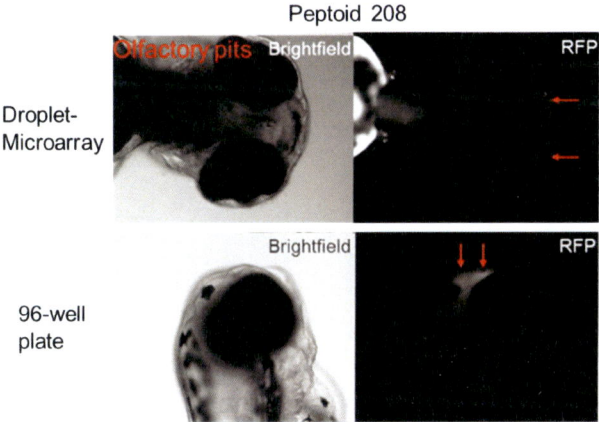

Figure 61: 72 hpf zebrafish incubated with P208 on Droplet-Microarrays in comparison to incubation in a 96-well plate. Images with brightfield and red filter (RFP) were taken with an Olympus Scan^R IX81 fluorescence microscope. Accumulation of peptoid was found for both method in the olfactory system of the zebrafish [203].

Furthermore, amount of peptoid, which was needed for the screening, was much lower for incubation in droplets. While 96-well-plate incubation needed 13.4 nmol peptoid, for DMA plates 1 nmol was sufficient to reach the same peptoid concentration in the fish water surrounding the embryo.

3.4. Influence of the dye

To investigate the localization of peptoids, in cells and zebrafish embryos, all peptoids in the different libraries were labeled with rhodamine B. However, the size of the dye compared to the peptoid, is relatively large, with almost equal molecular weights, and it would be possible, that rhodamine B has an effect on cellular uptake, intracellular localization and organ specificity. Since coupling of fluorescent dyes is indispensable for microscopic analyzes of peptoids, several representative peptoids were alternatively labeled with fluorescein (figure 62, 2) and a cyanine dye (figure 62, 3, synthesized by Alexander Braun, Institute of Organic Chemistry, KIT). All dyes have different structures and properties. Fluorescein represents a hydrophilic dye, whereas the cyanine dye is very lipophilic. Due to different properties of the dyes, octanol-water partition coefficients of peptoids change after coupling them to fluorescein and cyanine, which could lead to changes in cellular uptake.

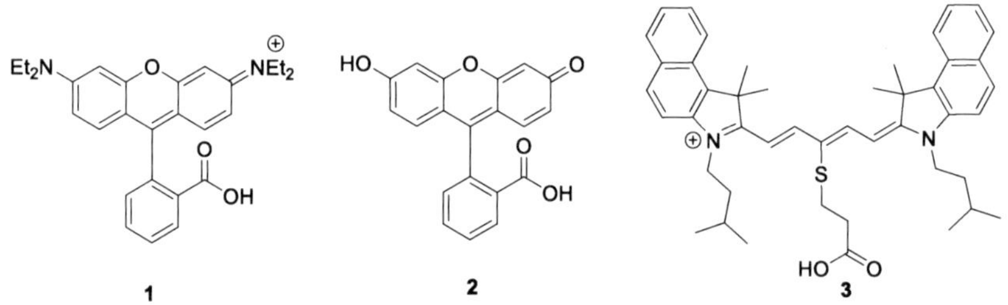

Figure 62: Dyes coupled to peptoids for investigation of the influence of fluorescent labeling on cellular uptake, intracellular localization and organ specificity. 1. Rhodamine B 2. Fluorescein 2. Cyanine dye.

For the investigation of the influence of the dye nine peptoid sequences were chosen, which displayed uptake in mitochondria or endosomes and had interesting, organ-specific localizations in zebrafish embryos. Peptoids were synthesized in IRORI MiniKans using the split-mix approach, as described in chapter 3.1.1.. Fluorescein labeled peptoids were synthesized in collaboration with Xenia Kempter (Bachelor thesis, Institute for Toxicology and Genetics). Sequences and molecular weights of peptoids synthesized with fluorescein and the cyanine dye are shown in table 12. With exception of P94-Fluo all peptoids were successfully synthesized and identified by mass spectrometry and purified by reversed-phase HPLC.

Table 12: Peptoid sequences coupled to fluorescein and cyanine dye. Peptoid numbers, respective sequence, dye and molecular weights are shown. Successful identified peptoids by mass spectrometry are marked with an "x".

Peptoid	Sequence	Dye	Molecular weight (g/mol)	Mass found (x)
P8-Fluo	Nprg-Nprg-Npcb-Nphe	Fluorescein	850.33	x
P31-Fluo	Nprg-Npcb-Nphe-Nlys	Fluorescein	883.40	x
P39-Fluo	Nprg-Nlys-Npcb-Nlys	Fluorescein	864.40	x
P94-Fluo	Npcb-Npcb-Nphe-Npcb	Fluorescein	1023.36	-
P170-Fluo	Nlys-Nlys-Nlys-Npcb	Fluorescein	897.47	x
P171-Fluo	Nlys-Nlys-Nlys-Nlys	Fluorescein	844.03	x
P172-Fluo	Nlys-Nlys-Nlys-Nphe	Fluorescein	863.03	x
P177-Fluo	Nlys-Nphe-Nprg-Nprg	Fluorescein	796.88	x
P194-Fluo	Nphe-Nprg-Nprg-Npcb	Fluorescein	850.33	x
P8-Cy	Nprg-Nprg-Npcb-Nphe	Cyanine	1218.03	x
P31-Cy	Nprg-Npcb-Nphe-Nlys	Cyanine	1251.11	x
P39-Cy	Nprg-Nlys-Npcb-Nlys	Cyanine	1232.10	x
P94-Cy	Npcb-Npcb-Nphe-Npcb	Cyanine	1391.07	x
P170-Cy	Nlys-Nlys-Nlys-Npcb	Cyanine	1265.18	x
P171-Cy	Nlys-Nlys-Nlys-Nlys	Cyanine	1211.73	x
P172-Cy	Nlys-Nlys-Nlys-Nphe	Cyanine	1230.74	x
P177-Cy	Nlys-Nphe-Nprg-Nprg	Cyanine	1164.59	x
P194-Cy	Nphe-Nprg-Nprg-Npcb	Cyanine	1218.03	x

For the analysis of cellular uptake and intracellular localization 1.5×10^4 HeLa cells were treated with 10 µM peptoid solution for 24 h. Subsequently, cells were incubated for 30 min with MitoTracker® Green (Rhodamine B and Cyanine labeled peptoids) or MitoTracker® Red (fluorescein labeled peptoids) to stain mitochondria, washed with DPBS and cell nuclei were stained with Hoechst 33342. Cells were investigated by confocal microscopy and images of P8 (Fluo, RhodB, Cy), P194 (Fluo, RhodB, Cy), P31 (Fluo, RhodB, Cy) and P170 (Fluo, RhodB, Cy) are shown in figure 63 and figure 64. P8 and P194 represent two MPPos, as they displayed strong mitochondrial localization in cell screening of library 1. P8-Cy and P194-Cy showed the same intracellular localization as they are clearly overlapping with the MitoTracker® Green signal. P8-Fluo and P194-Fluo were also penetrating the cell membrane and were accumulating in the cell. However, only low mitochondrial uptake was found as they were mainly localized in the cytosol. An explanation for this change in uptake could be the hydrophilicity of fluorescein. To cross the mitochondrial membranes certain lipophilicity of compounds is needed. For the hydrophilic peptoids P31 and P170 no difference in uptake and intracellular localization is visible. P31-Fluo, P31-RhodB and P31-Cy all displays strong endosomal uptake. P170-Fluo, P170-RhodB and P170-Cy were also all located in endosomes. However, the uptake was lower, which was probably due to the high amount of Nlys. As it has been shown in the automated evaluation of peptoid uptake (chapter 3.1.1.) an increasing amount of Nlys led to a decrease of peptoid uptake.

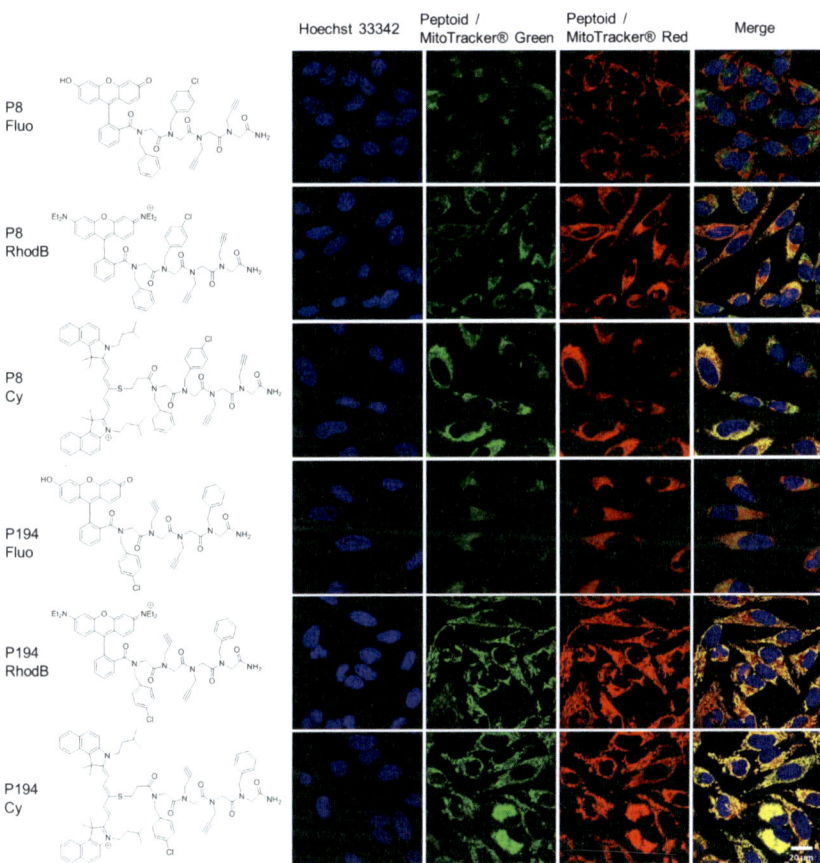

Figure 63: Sequences and microscopic images of cellular uptake of P8-Fluo, P8-RhodB, P8-Cy, P194-Fluo, P194-RhodB and P194-Cy in HeLa cells. 1.5 x 10⁴ cells were treated with 10 µM peptoid for 24 h at 37 °C. Accumulation of peptoids was detected with fluorescence confocal microscopy Leica TCS-SPE, (Objective: ACS APO 63x/1.30 OIL). 1. Hoechst 33342, Ex.: 405 nm, Em.: 417-468 nm, 2. MitoTracker® Green Ex.: 488 nm, Em.: 499-552 nm 3. Peptoid, Fluorescein: Ex.: 488 nm, Em.: 490-590 nm, Rhodamin B: Ex.: 560 nm, Em.: 590 - 690 nm, Cyanine dye: Ex.: 633 nm, Em.: 640- 710 nm 4. Merge, Scale bar: 20 µm

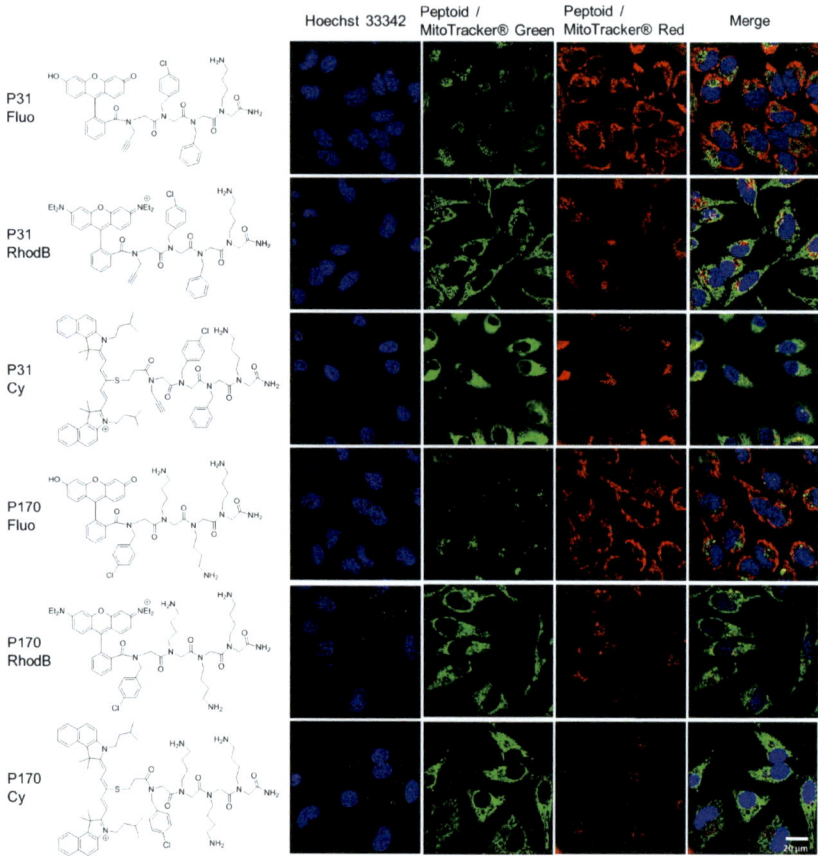

Figure 64: Sequences and microscopic images of cellular uptake of P31-Fluo, P31-RhodB, P31-Cy, P170-Fluo, P170-RhodB and P170-Cy in HeLa cells. 1.5 x 10^4 cells were treated with 10 µM peptoid for 24 h at 37 °C. Accumulation of peptoids was detected with fluorescence confocal microscopy Leica TCS-SPE, (Objective: ACS APO 63x/1.30 OIL).). 1. Hoechst 33342, Ex.: 405 nm, Em.: 417-468 nm, 2. MitoTracker® Green Ex.: 488 nm, Em.: 499-552 nm 3. Peptoid, Fluorescein: Ex.: 488 nm, Em.: 490-590 nm, Rhodamin B: Ex.: 560 nm, Em.: 590 - 690 nm, Cyanine dye: Ex.: 633 nm, Em.: 640- 710 nm 4. Merge, Scale bar: 20 µm

A summary of intracellular localizations of all peptoids is shown in table 13. The influence of the dye on cellular uptake and intracellular accumulation seemed to be low. The cyanine dye and fluorescein also displayed different colocalization for hydrophilic peptoids in comparison to hydrophobic peptoids. Like the hydrophilic rhodamine B labeled peptoids, all differently labeled hydrophilic peptoids, were also accumulating in endosomes. For the differently labeled hydrophobic peptoids no differences between rhodamine B and cyanine labeled peptoids could be distinguished. However, fluorescein peptoids, were mainly localized in the cytosol. It can be concluded, that fluorescent labeling is a suitable approach for cell-screenings of CPPos.

Table 13: Peptoids, coupled to fluorescein, rhodamine B and cyanine dye and their respective intracellular localization in HeLa cells.

Peptoid	Fluorescein	Rhodamine B	Cyanine
P8	cytosol	mitochondria	mitochondria
P31	endosomes	endosomes	endosomes
P39	endosomes	endosomes	endosomes
P94	-	mitochondria	mitochondria
P170	endosomes	endosomes	endosomes
P171	endosomes	endosomes	endosomes
P172	endosomes	endosomes	endosomes
P177	endosomes	endosomes	endosomes
P194	cytosol	mitochondria	mitochondria

Furthermore, uptake of peptoids was not only compared in cells but also organ specificity was compared by testing peptoids in zebrafish embryos. It is possible, that already small changes in structure and properties of peptoids could have much more influence on complex organisms compared to cells. Therefore, 4 hpf zebrafish embryos were incubated for 2 h with 50 µM peptoid solution in fish water. Subsequently, zebrafish were washed with fish water, anesthetized with tricaine, and analyzed by confocal microscopy. Figure 65 shows the comparison of P8-Fluo with P8-Cy and P171-Fluo with P171-Cy. In the embryo screening of the rhodamine B labeled peptoids from library 1, the hydrophobic P8-RhodB was localized in the olfactory system and the kidney. In contrast, the hydrophilic peptoid P171-RhodB was accumulating in the cardinal vein. Unfortunately, these localizations were not found for their fluorescein and cyanine labeled peptoid- counterparts. P8-Cy was accumulating mainly in skin cells of zebrafish embryos. Additionally, it was found in the gills in the region of the olfactory system. Uptake of P8-Fluo was relatively low and the peptoid could only be detected in the stomach. The hydrophilic peptoid P171-Cy was also accumulating in skin cells and gills. For the P171-Fluo localization was found in the stomach and in the liver.

	Brightfield	Peptoid	Merge
P8-Cy			
P8-Fluo			
P171-Cy			
P171-Fluo			

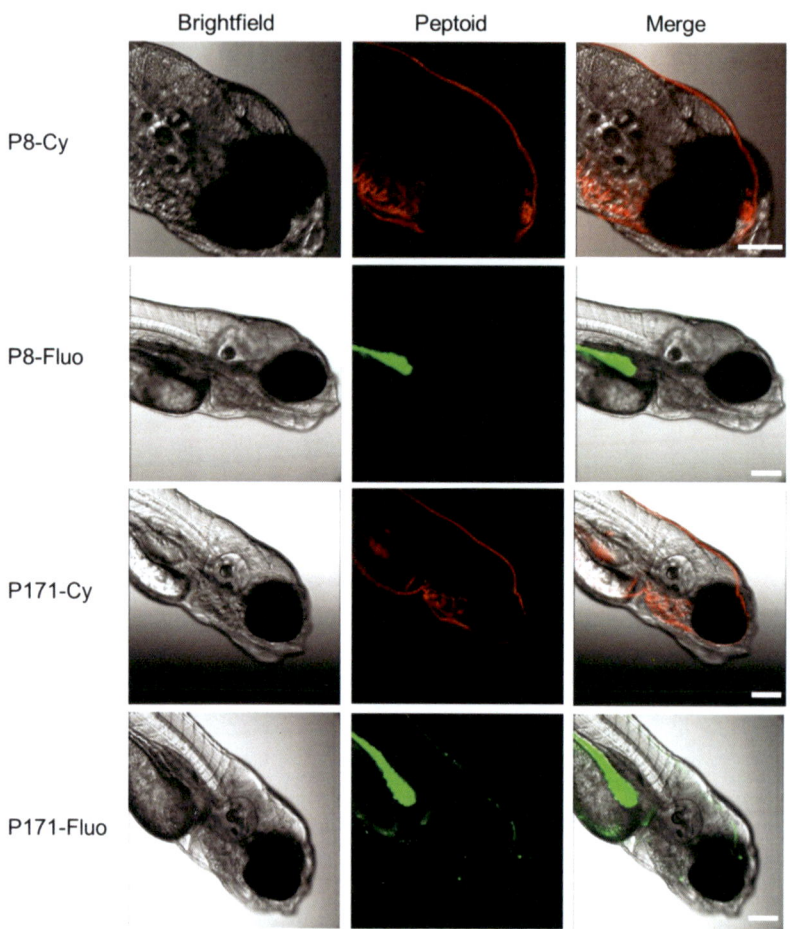

Figure 65: 4 dpf zebrafish incubated for 2 h with P8-Cy, P8-Fluo, P171-Cy and P171-Fluo, Leica TCS-SPE, objective: ACS APO 10x/0.30 DRY. 1. Peptoid, Cyanine labeled Peptoids Fluorescein: Ex.: 488 nm, Em.: 490-590 nm, Cyanine dye: Ex.: 633 nm, Em.: 640- 710 nm 2. Brightfield 3. Merge, Scale bar: 150 µm

Similar results were found for the other seven peptoid sequences. All fluorescein labeled peptoids were accumulating exclusively in the stomach and cyanine labeled peptoids were accumulating mainly in the skin and the gills. A summary for all localizations is shown in table 14. Furthermore, it was found, that toxicity seems to differ for different dyes. Zebrafish embryos which were incubated with rhodamine B labeled peptoid solutions for more than 2 h displayed toxic phenotypes, e.g. apoptotic cells an inhibition of movements. Overnight

108

incubation usually led to the death of the embryos. In contrast, no increase of uptake or toxic effects for fluorescein and cyanine labeled peptoids was visible after overnight incubation.

Table 14: Fluorescein, rhodamine B and cyanine dye coupled peptoids and their localization in 4 dpf zebrafish larvae

Peptoid	Fluorescein	Rhodamine B	Cyanine
P8	stomach	olfactory system, kidney	ckin, gills, olfactory system/nasal cavity
P31	stomach	neuromast cells	skin, gills, olfactory system/nasal cavity
P39	stomach	neuromast cells	skin, gills
P94	stomach	olfactory system	skin, gills
P170	stomach	caudal vein	skin, gills
P171	stomach, liver	caudal vein	skin, gills
P172	stomach	neuromast cells	skin, gills,
P177	stomach	stomach	skin, gills, olfactory system/nasal cavity
P194	stomach	olfactory system, kidney	skin, gills

For further investigation of potential localization of P8-Cy, P31-Cy and P177-Cy in the olfactory system peptoids were tested in Runx:GFP zebrafish embryos. The Runx:GFP fish lines expresses the Runx-protein fused to GFP in the heart and the olfactory system. An overlapping peptoid signal with Runx-GFP-protein signal could prove accumulation of peptoids in the olfactory system. This experiment was already performed by Franziska Rönicke to confirm localization of rhodamine B labeled peptoids [91]. F. Rönicke could show, that rhodamine B labeled peptoids accumulate in the olfactory neurons [91]. To analyze accumulation of cyanine labeled peptoids, zebrafish embryos were incubated with 50 µM peptoid solution overnight, washed with fish water and analyzed by confocal microscopy. Overview of uptake in upper part of zebrafish as well as a magnified view of the olfactory system is shown in figure 66. Even though a high colocalization of the GFP signal with the peptoid signal was found in overview image, the colocalization in the magnified view was rather low.

Brightfield	GFP	Peptoid	Merge

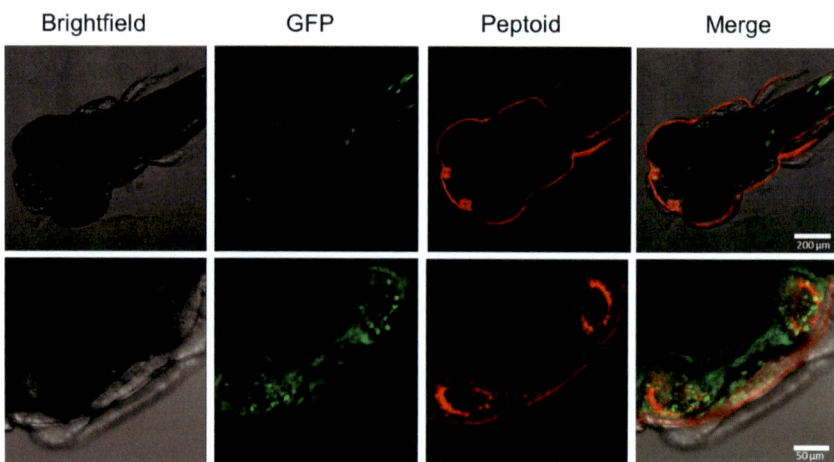

Figure 66: 4 dpf Runx:GFP zebrafish incubated for 2 h with P8-Cy, Leica TCS-SPE, Objective: ACS APO 10x/0.30 DRY. 1. Brightfield 2. GFP, Ex.: 488 nm, Em.: 490-590 nm 3. Peptoid, Ex.: 633 nm, Em.: 640-710 nm 4. Merge, Scale bar: 200 µm and 50 µm

Z-stack images of the olfactory system and comparison of rhodamine B labeled peptoid P116-RhodB (*N*pcb-*N*phe-*N*prg-*N*phe-RhodB) with cyanine labeled peptoid P8-Cy (*N*prg-*N*prg-*N*pcb-*N*phe-Cy), shown in figure 67, displayed higher colocalization of the rhodamine B peptoid compared to the cyanine labeled peptoid with the Runx:GFP signal. It might be possible that cyanine labeled peptoids accumulated mainly in skin cells of the nasal cavity. These findings showed, that fluorescent labeling might have an effect on organ specificity of peptoids. Due to the high lipophilicity of the cyanine dye labeled peptoids might not be able to penetrate the skin of zebrafish anymore, while fluorescein labeled peptoids accumulate mainly in the stomach, suggesting uptake by swallowing of surrounding peptoid solution.

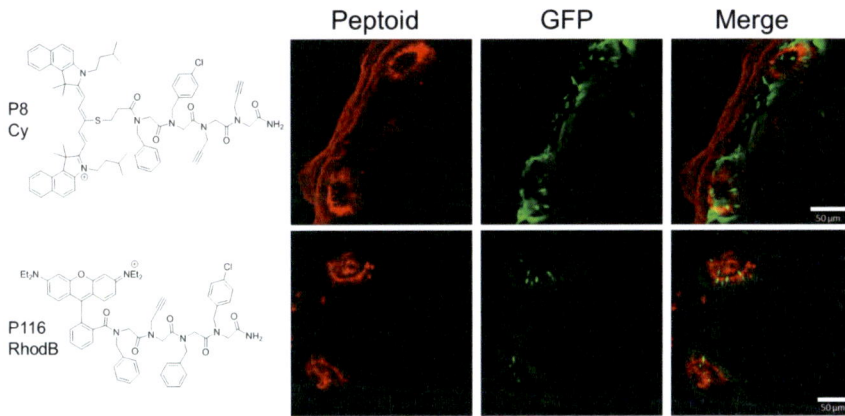

Figure 67: 4 dpf Runx:GFP zebrafish incubated for 2 h with P8-Cy and P116-RhodB, Leica TCS-SPE, Objective: ACS APO 10x/0.30 DRY. 1. Peptoid, RhodB: Ex.: 560 nm, Em.: 590-690 nm, Cyanine: Ex.: 633 nm, Em.: 640- 710 nm 2. GFP, Ex.: 488 nm, Em.: 490-590 nm 3. Merge, Scale bar: 50 μm

3.4.1. TADF Transporters

Not only peptoids with common dyes like rhodamine B and fluorescein were analyzed for biological application but also peptoids linked to TADF dyes (Thermally Activated Delayed Fluorescence), in cooperation with Stephan Münch and Fabian Hundemer (Institute of Organic Chemistry, KIT). Lately, TADF dyes became highly interesting due to their potential application in organic light emitting diodes (OLEDs). Fluorescent molecules which are commonly used for the light emitting layer of OLED have low fluorescence efficiency, whereas TADF dyes could be an alternative to overcome this problem [205]. Classical fluorescent compounds cannot convert 100% of electrical excitation energy in emission as they lose around 75% energy due to *intersystem crossing* from singlet state to triplet state (S_1 -> T_1). TADF dyes show *reverse intersystem crossing* (T_1 -> S_1) and therefore display delayed fluorescence [206]. However, TADF dyes are not only investigated in material science, but were also recently analyzed for biological applications. To increase water-solubility and cell-penetration abilities, alkyne functionalized TADF dyes were coupled to azide functionalized cell penetrating peptoids *via* 1,3 dipolar cycloaddition, by S. Münch and F. Hundemer [207]. Two different heptameric peptoids were synthesized by submonomer based method with an N-terminal azide submonomer (*N*-azidopropylamine). While TADF transporter 1 is hydrophilic as it consisted only of *N*lys side chains (*N*lys-*N*lys-*N*lys-*N*lys-*N*lys-*N*lys-N4az), TADF transporter 2 contained

111

alternating *N*phe and *N*lys side chains (*N*lys-*N*phe-*N*lys-*N*phe-*N*lys-*N*phe-N4az) Afterwards, TADF was coupled to peptoids by copper catalyzed 1,-3- alkyne-azide cycloaddition (scheme 6).

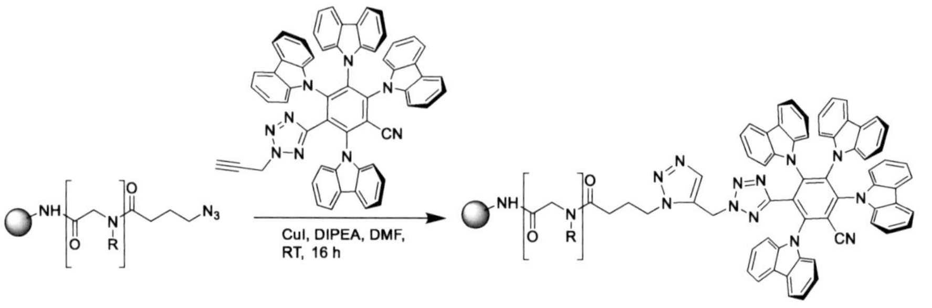

Scheme 6: Synthesis of TADF labeled peptoids *via* CuAAc [207].

The TADF transporters were purified by HPLC and identified by mass spectrometry. The structure of Transporter 1 and Transporter 2 is shown in figure 68.

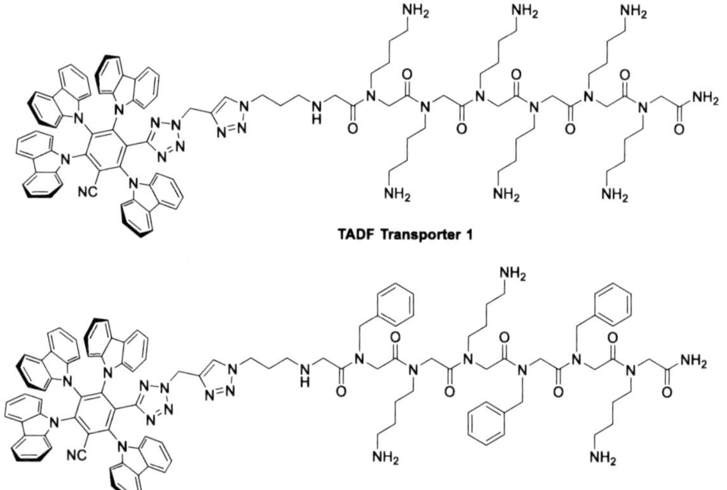

Figure 68: TADF transporter 1 (*N*lys-*N*lys-*N*lys-*N*lys-*N*lys-*N*lys-TADF) and TADF transporter 2 (*N*lys-*N*phe-*N*lys-*N*phe-*N*lys-*N*phe-TADF) [207].

Subsequently, the TADF-transporters 1 and 2 were analyzed for biological application in different cell lines and zebrafish embryos. 1×10^4 HeLa and SK-MEL-28 cells and 2×10^4 RAW cells were incubated with 10 μM of respective TADF transporter for 24 h. After several washing steps with DPBS cellular uptake and intracellular localization was investigated by fluorescent confocal microscopy (figure 69). Transporters were excited with UV light (405 nm) and strong emission signals could be detected. Transporter molecules were taken up by HeLa, SK-MEL-28 and RAW cells and were detected mainly in endosomes. This is probably due to the positive lysine side chain, which is present in both peptoids. Although transporter 2 is more lipophilic compared to transporter 1, no differences in uptake and localization could be observed. Another reason for endosomal uptake could be due to the large aromatic TADF dyes and possible π-interactions between the dyes, hindering the ability for direct membrane penetration of peptoids. In addition, no differences between the uptake in different cell types, such as HeLa, SK-MEL 28 and RAW cells were determined.

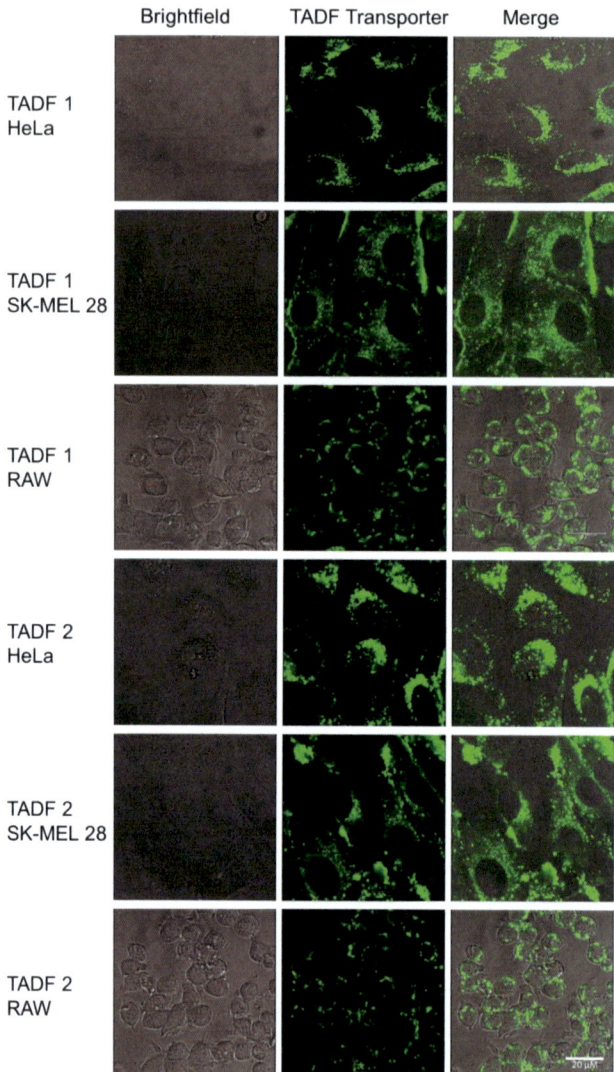

Figure 69: Cellular uptake of transporter 1 and 2 in HeLa, SK-MEL 28 and RAW cells. Cells were treated with 10 µM transporter for 24 h at 37 °C. Accumulation of transporters was detected with fluorescence confocal microscopy Leica TCS-SPE, Objective: ACS APO 63x/1.30 OIL). 1. Brightfield 2. TADF 1 or TADF 2, Ex.: 405 nm, Em.: 420-550 nm 3. Merge, Scale bar: 20 µm

For further investigation of fluorescence and optimal detection of TADF labeled peptoids, emission spectra of TADF was measured in living cells (HeLa) by performing a lambda scan (figure 70). Therefore, the transporters were excited with 405 nm and a series of images with different emission wavelengths were taken and fluorescence intensity was measured for each image. Emission was found between 410 and 550 nm with a maximum of 480 nm. Additionally, cellular uptake was analyzed for different transporter concentrations (0.5 µM, 1 µM, 5 µM and 10 µM) by confocal imaging and subsequently measuring fluorescence intensity of single cells for each concentration with ImageJ. Cells displayed increasing uptake of the transporters with increased concentrations and uptake was found to be nearly exponential (figure 70).

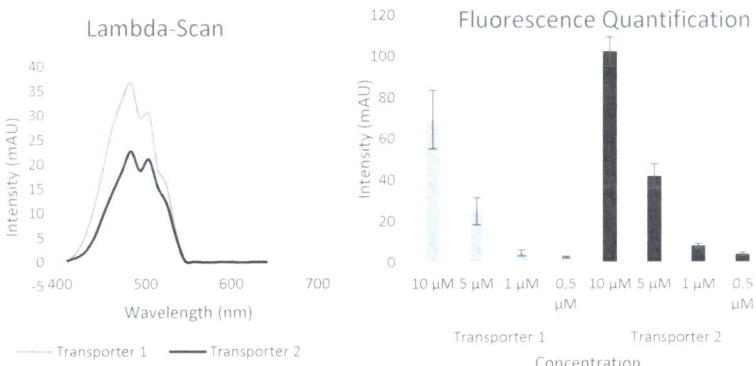

Figure 70: Lambda-Scan (left) and fluorescence quantification (right) of TADF transporter 1 and transporter 2 in HeLa cells.

Another important aspect for the biological application of TADF dyes is their cytotoxicity. Hence, MTT assays with HeLa cells were performed for concentrations in low micromolar range. Cells were incubated for 72 h with 1, 5, 10, 20 and 30 µM peptoid solution. In this range both transporters had no toxic effects on cells and LD_{50} values could not be determined (figure 71). These findings confirm biocompatibility of TADF dyes.

115

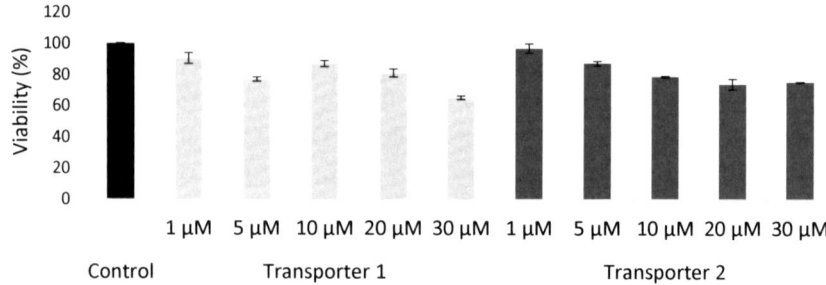

Figure 71: Cytotoxicity assay of transporter 1 and transporter 2 in HeLa cells. Transporters were tested in 1, 5, 10, 20 and 30 μM concentration for 72 h and viability was determined using the MTT assay.

Transporter 1 and transporter 2 were not only analyzed in cells but also tested for organ specificity in zebrafish embryos. 4 dpf zebrafish were incubated for 2 h with 20 μM transporter in E3 medium in the dark. Afterwards, they were washed several times with medium, anesthetized with tricaine and the uptake was investigated by confocal microscopy. However, no specific organ localization could be detected and the transporters 1 and 2 were mostly detected in gills, stomach and skin cells. Furthermore, transporters displayed toxicity in fish and several embryos died during the incubation time. Due to this effect, bathing of zebrafish might not be a suitable application for TADF labelled peptoids. To improve transporter uptake and simultaneously lower the toxicity, 2 dpf zebrafish were anesthetized with tricaine and injected with 1 mM transporter 1 and 2. Afterwards, the embryos were incubated in fish water for 24 h in the dark at 28 °C. Finally, fish were anesthetized again and uptake was investigated by confocal microscopy (figure 72). Injected fish displayed no obvious toxic effects and transporters were found to accumulate mainly in endothelial cells and macrophages of the blood vessels. In addition, they could be detected in the brain ventricle. The fluorescence signal detected from the yolk sac, depicted in figure 72, was not due to the TADF transporters, but to autofluorescence, which was also found in the control group.

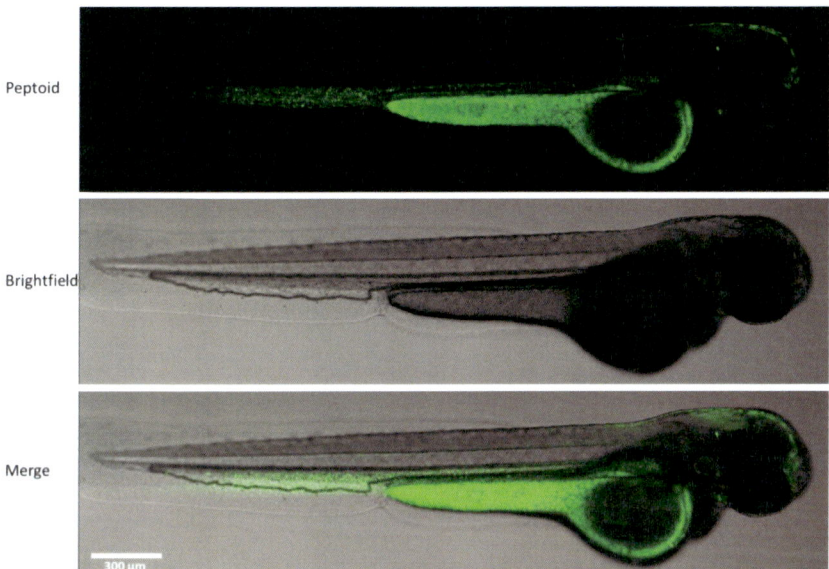

Figure 72: 3 dpf old zebrafish injected with 1 mM transporter 1. Accumulation of transporter 1 was detected with fluorescence confocal microscopy Leica TCS-SPE, Objective: ACS APO 10x/0.30 DRY. 1. Transporter 1, Ex.: 405 nm, Em.: 420-550 nm 2. Brightfield 3. Merge, Scale bar: 300 μm

3.5.   Single chain nanoparticles for drug delivery

Lately nanotechnology and nanoscience became a growing medical research field, as there are many different applications for nanoparticles. A wide variety of substances has been investigated for the formation of nanoparticles, for example metals, polymers or biological materials, such as phospholipids and lipids [208]. Their small size allows them to penetrate cell membranes. Sometimes they are even able to cross the nuclear membrane or accumulate in different organelles [209]. The advantages of nanoparticles are not only their long circulation in the blood system and reduced clearance by the kidney, leading to an enhanced bioavailability, but also their large functional surface [210, 211]. Due to their large surface to mass ratio various surface modifications can be made to achieve organ-specific accumulation and to bind and adsorb several different kinds of drugs. A challenging task in drug delivery is the application of hydrophobic drugs as they usually must be applied in high concentrations for sufficient uptake and lead to gastrointestinal mucosal toxicity [212]. To overcome this issue nanoparticles have been successfully used as it has been shown for the treatment with anticancer drugs such as paclitaxel by binding to albumin-nanoparticles [213]. Promising results were also found for PK1, a polymer-doxorubicin conjugate for the treatment of cancer, which showed significantly better results in comparison to doxorubicin alone [214]. It is assumed, that nanoparticles prevent the discarding of drugs by efflux transporters and are therefore highly interesting in terms of multi-drug-resistant cells [215, 216]. It has also been shown that nanoparticles are not only able to target drugs to specific tissues but are also able to cross various barriers within the body, e.g. the BBB. Therefore, fluorescent, water-soluble polymer nanoparticles were investigated for drug delivery *in vitro* and *in vivo*. The so-called single chain polymer nanoparticles (SCNPs) were based on poly(acrylic acid) (PAA), which is known to be biocompatible and non-toxic [217]. Synthesis of SCNPs was carried out by Carolin Heiler (Institute for Chemical Technology and Polymer Chemistry, KIT) [217]. A homopolymerization of acrylic acid was performed *via* Reversible Addition Fragmentation Chain Transfer (RAFT) polymerization. Subsequently, the water-soluble linear poly acrylic acid was partially functionalized with photoreactive compounds such as tetrazole (Tet) and furan-protected maleimide (p-Mal) as well as the non-photoreactive tetra ethyleneglycol monomethylether (TEG) for increasing the bioavailability (PAA$_{90}$(Tet/p-Mal/TEG)). The tetrazole and furan-protected maleimide (p-Mal) were folded *via* dual nitrile imine-mediated tetrazole-ene cycloaddition (NITEC) [217, 218]. A solution of the polymer precursor (17 mg/L, in deionized water) was prepared and irradiated with UV-light (λ_{max}=320 nm). The resulting SCNPs display a size of 2.4 nm and exhibit fluorescence in the 570 nm regime. A schematic representation of the formation is shown in figure 73.

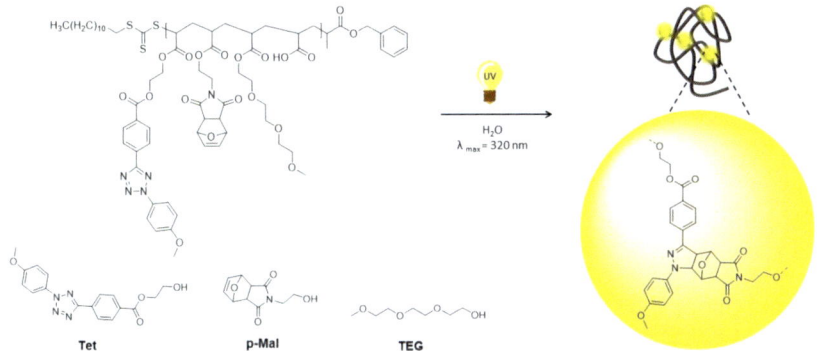

Figure 73: Schematic representation of the SCNP formation. Functional poly acrylic acid polymer bearing photoreactive (Tet and p-Mal) as well as (TEG) moieties were irradiated with UV-light in pure water [217, 218].

To investigate SCNPs for drug delivery, cell penetration abilities and intracellular accumulation were tested in HeLa, SK-MEL-28, RAW cells and HUVEC cells. 1.5 x 10^4 cells were treated with 150 µg/ml SCNPs for 24 h at 37°C. Afterwards cells were washed with DPBS and cellular uptake was investigated by confocal microscopy (figure 74). SCNPs were excited with 405 nm and emission was measured between 450 and 550 nm. A high fluorescent signal was found for all investigated cell lines after excitation at 405 nm, indicating that nanoparticles were able penetrate the cell membranes. SCNPs were mainly accumulating in the cytosol and endoplasmic reticulum. Differences in uptake or localisation for the different tested cell lines could not be determined. Accumulation in the cytosol is of high interest for the delivery of various drugs such as siRNA. It has been shown for many other nanoparticles that they accumulate mainly in endosomes or lysosomes and endosomal escape is hard to achieve. A possible reason for cytoplasmic accumulation of SCNPs could be their small size compared to other investigated nanoparticles.

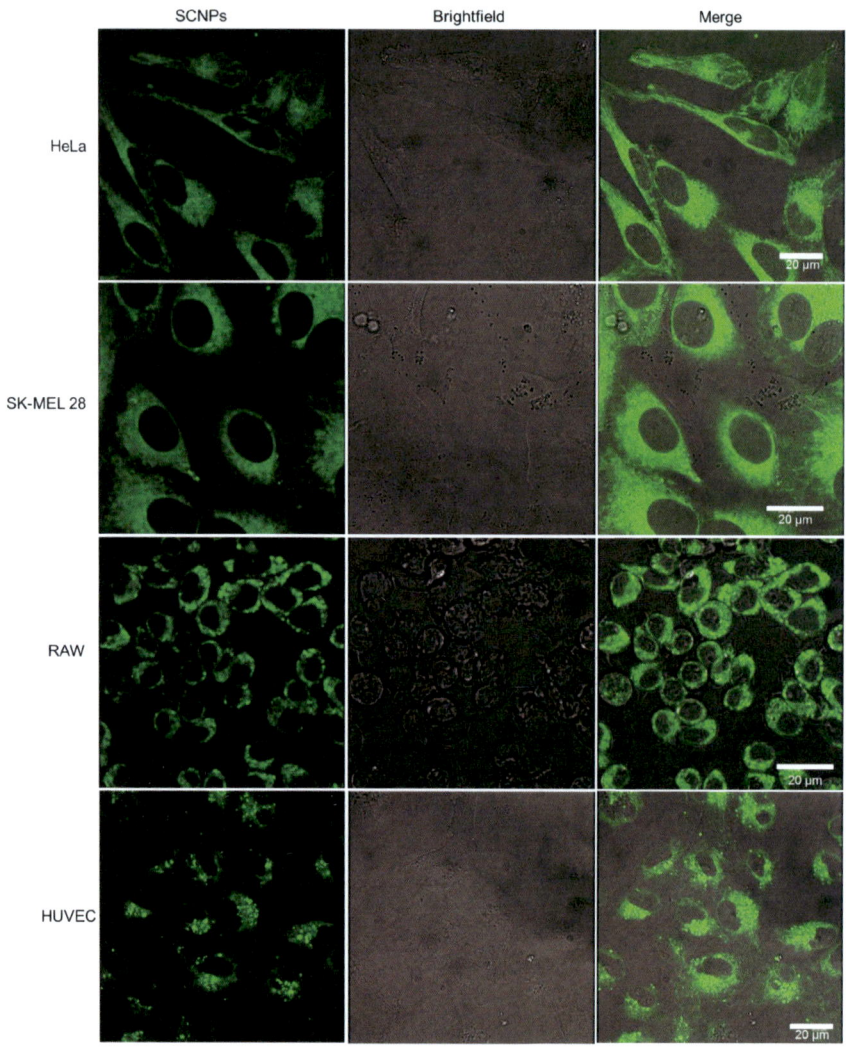

Figure 74: Cellular uptake of SCNPs in HeLa, SK-MEL 28, RAW and HUVEC cells. 1.5 x 10^4 cells were treated with 150 µg/ml SCNPs for 24 h at 37 °C. Accumulation of SCNPs was detected with fluorescence confocal microscopy Leica TCS-SPE, Objective: ACS APO 63x/1.30 OIL). 1. SCNPs, Ex.: 405nm, Em.: 450-550 nm 2. Brightfield 3. Merge, Scale bar: 20 µm

A huge drawback of nanoparticles in pharmaceutical research is their high cytotoxicity *in vitro* and *in vivo* [219-221]. Therefore, SCNPs were analyzed in HeLa cells, red blood cells and zebrafish embryos to investigate their effects in cells and whole organisms. Concentrations between 15 and 225 µg/ml SCNPs were tested for cytotoxicity in HeLa cells (figure 75).

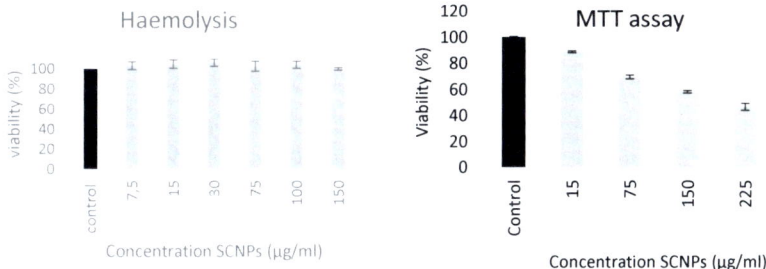

Figure 75: Toxicity tests of SCNPs. Left: Haemolysis of red blood cells of rats treated with SCNPs (7.5, 15, 30, 75, 100 and 150 µg/ml) for 1 h. Right: Cytotoxicity of SCNPs (15, 75, 150, 225 µg/ml) in HeLa cells after 72 h, viability was measured after incubation with MTT.

SCNPs displayed a moderate toxicity with and an LD_{50} value around 200 µg/ml. Haemolysis was tested in red blood cells of rats with concentrations between 7.5 and 150 µg/ml (figure 75). For red blood cells no toxicity was determined and LD_{50} was found to be higher than 150 µg/ml concentration. The cytotoxicity in zebrafish embryos was tested using the fish embryo toxicity test (FET) as an alternative to the fish acute toxicity test. Therefore, fertilized zebrafish eggs (4 hpf) were incubated with different concentrations of SCNPs between 7.5 and 100 µg/ml. Zebrafish were incubated at 27 °C for four days and analyzed each day by stereo microscopy. The mortality rate was relatively low, LD_{50} for zebrafish embryos could not be determined and zebrafish embryos were developing normal for concentrations between 7.5 and 30 µg/ml. However, for higher concentrations differences between treated and control fish could be monitored. For those embryos the hatching rate was significantly decreased and even 96 hpf about 30% of zebrafish incubated with 100 µg/ml did not hatch. Delayed hatching rate could be due to inhibition of movements of embryos or caused by strong accumulation of SCNPs in the chorion. For one embryo incubated with 100 µg/ml SCNPs deformation of body was visible. Developed eyes were smaller and body composition was different to control fish. However, most of the hatched embryos had developed normal and showed no visual difference to the control.

Table 15: FET test for SCNPS: analysis of influence on the development of zebrafish larvae after incubation with SCNPs (7.5, 15, 30, 75 and 100 µg/ml). Zebrafish embryos were investigated with stereo microscopy after 24, 48, 72 and 96 h.

Concentration SCNPs (µg/ml)	7.5	15	30	75	100	Control
Mortality (%)	0	0	0	14	0	0
Deformation (%)	0	0	0	0	14	0
Degradation of movements	no	no	no	yes	yes	no
Hatching rate (48 h)	57	0	0	0	0	43
Hatching rate (72 h)	100	100	100	71	57	100
Hatching rate (96 h)	100	100	100	83	71	100

Furthermore, SCNPs were investigated for potential organ specificity, biocompatibility, stability and distribution *in vivo*. Due to strong fluorescence of SCNPs and high transparency of zebrafish embryos accumulation of particles could be analyzed. Naturally hatched, 96 hpf casper zebrafish embryos were incubated with 150 µg/ml SCNPs in E3-medium for 2 h and 24 h. Subsequently, zebrafish were washed with E3-medium and analyzed by confocal microscopy. SCNPs had no visible toxic effects on larvae and uptake of SCNPs was very low and no clear organ specificity could be determined. However, low accumulation was found in stomach and digestive system, indicating uptake by ingestion. To increase the uptake SCNPs were injected in cardinal vein of dechorionated, anesthetized, 48 hpf zebrafish embryos. Afterwards zebrafish were incubated at 27 °C for 24 h and following analyzed by confocal microscopy. All injected zebrafish recovered normally after injection and accumulation was found mainly in endothelial cells and macrophages.

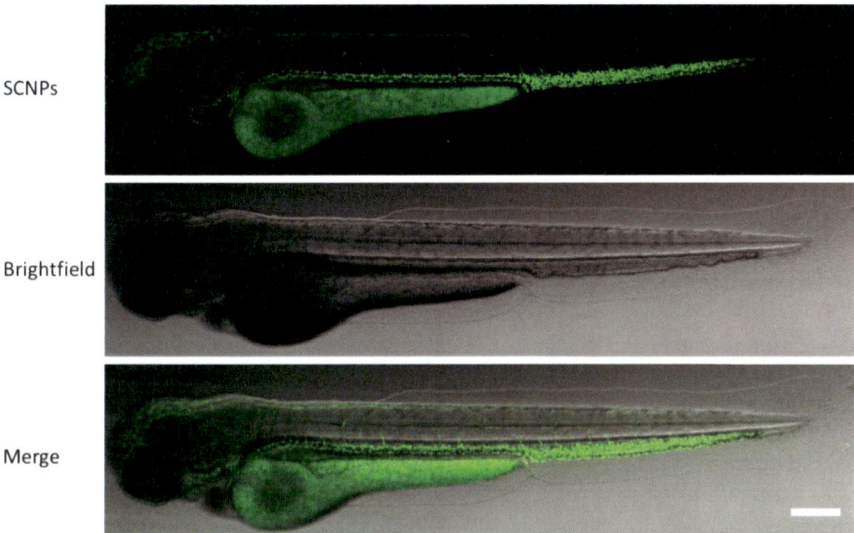

SCNPs

Brightfield

Merge

Figure 76: 24 h after caudal vein injections of 15 mg/ml SCNPs A: Leica TCS-SPE, objective: ACS APO 10x/0.30 DRY. 1. SCNPs, Ex.: 405nm, Em.: 450-550 nm 2. Brightfield 3. Merge B: Tail, Leica TCS-SPE, Objective: 20x DRY. 1. SCNPs, Ex.: 405nm, Em.: 450-550 nm 2. Brightfield 3. Merge, Scale bar: 200 µm

The organ system of zebrafish as well as the BBB can be compared to the organ organization in humans. Therefore, zebrafish embryos are optimal to study brain specific delivery of nanoparticles. Potential brain specific uptake of SCNPs was analyzed by injection in the brain ventricles of 48 hpf dechorionated and anesthetized zebrafish embryos. Success of injection could be controlled directly after injection by microscopy as SCNPs were highly fluorescent and the accumulation in the brain ventricle was clearly visible. After injection embryos were embedded in 1.5% low-melting agarose and brain specific uptake was analyzed by confocal microscopy. Accumulation of SCNPs was investigated for a time range of 1 – 4 h after injection and after 24 h. Between 1 h and 3 h after injection accumulation of SCNPs was not only found in brain ventricle but also in brain tissue (figure 77). Localization of SCNPs was analyzed again after 24 h showing that SCNPs were accumulating in endothelial cells or recognized by macrophages in the blood system.

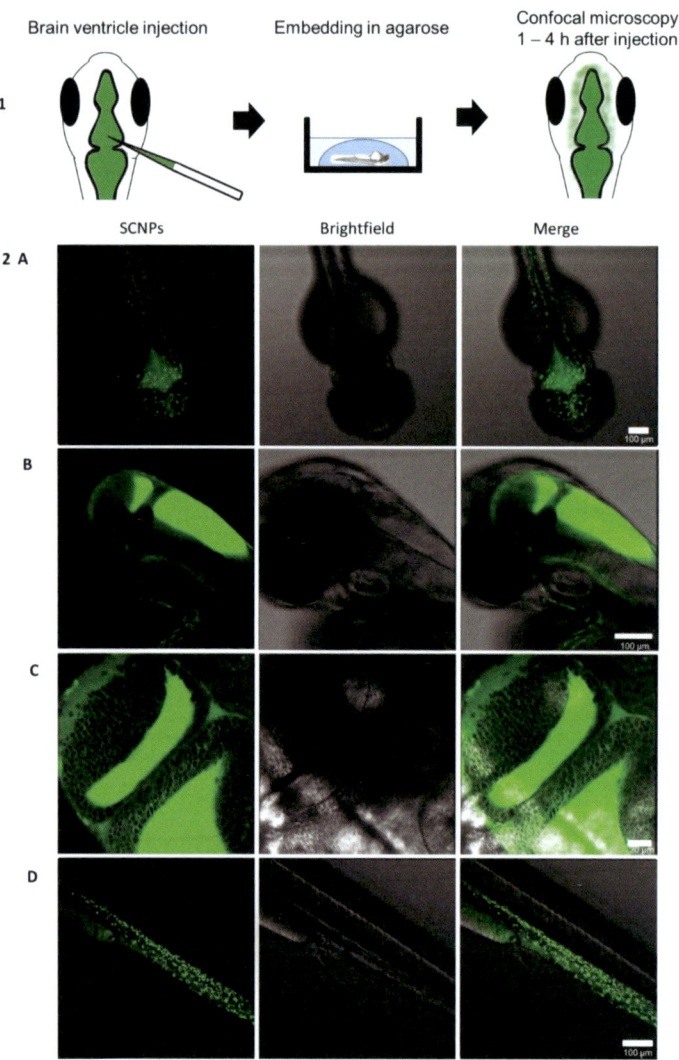

Figure 77: 2 dpf old zebrafish, 1 h after ventricle injections of 15 mg/ml SCNPs A: Dorsal view of brain ventricle, Leica TCS-SPE, objective: ACS APO 10x/0.30 DRY. 1. SCNPs, Ex.: 405nm, Em.: 450-550 nm 2. Brightfield 3. Merge B: B and C: magnified view of brain ventricle and brain, Leica TCS-SPE, Objective: 20x DRY. 1. SCNPs, Ex.: 405nm, Em.: 450-550 nm 2. Brightfield 3. Merge D: Tail, Leica TCS-SPE, Objective: 10x DRY. 1. SCNPs, Ex.: 405nm, Em.: 450-550 nm 2. Brightfield 3. Merge

4.   Summary and conclusion

Four different peptoid libraries, containing tetrameric, hexameric, decameric and dodecameric peptoids with a mixture of hydrophilic and hydrophobic side chains were analyzed for their suitability as organelle- and organ-specific transporter molecules. Peptoids were analyzed for cellular uptake, intracellular localization, cytotoxicity and organ specificity, using the zebrafish as a model organism. The fully permutated library 1, containing Nlys, Nprg, Nphe and Npcb (256 peptoids) was analyzed by an automated evaluation process for identification of a new class of mitochondria-penetrating peptoids (MPPos). The evaluation was based on colocalization of the MitoTracker® Green signal with the peptoid signal and revealed 72% endosomal peptoids and 26% mitochondrial peptoids with a success rate of 96%. MPPos are especially interesting for drug delivery, due to the importance of mitochondria for cellular functions and their potential role in diseases, and were therefore further analyzed determining their octanol-water-partition coefficients and cytotoxicity. MPPos are highly lipophilic, with LogP values ~2.5 (experimental), making them to interesting candidates for delivery across the BBB.

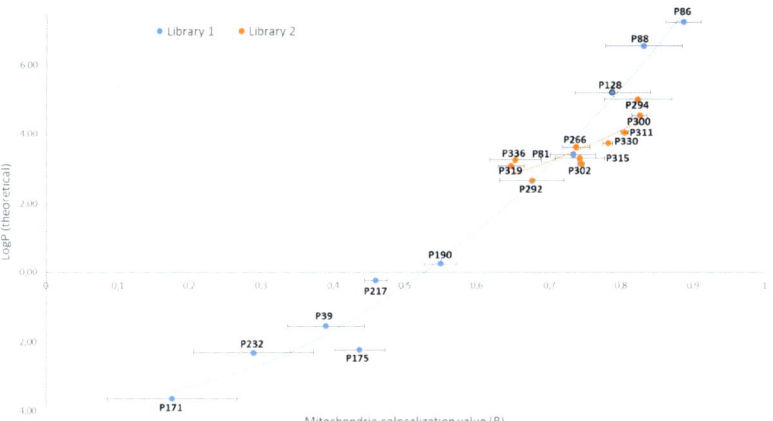

Figure 78: Colocalization values of peptoid signal with MitoTracker® Green signal, calculated with ImageJ, as a function of their LogP value, calculated with *Molinspiration*. 10 representative peptoids are shown for library 1 (blue) and 10 representative peptoids for library 2 (orange).

Furthermore, there was a strong connection between the LogP value of a peptoid and its intracellular localization. While hydrophilic peptoids displayed low colocalization values with the MitoTracker® Green, the value was increasing with rising LogP values, as shown in figure 78 (blue). These findings confirm the similarity of the cell penetrating peptoids to their peptide counterparts, as Horton et al. could already show similar correlations for peptides [222].

The second evaluated tetrameric peptoid library was more lipophilic, containing a full permutation of Npob, Nphe and Npcb residues. Due to the decreased hydrophilicity of peptoids in this library the amount of MPPos was increased (66%). Only high ratios of the side chain Npob, mimicking the natural, polar amino acid tyrosine, were inducing endosomal uptake of the peptoid. As library 2 contained no hydrophilic side chains, LogP values were less distributed (between ~2 and ~4). Calculated data for LogP values as a function of the mitochondria colocalization values were in agreement with the data, found for library 1, as shown in figure 78 (orange).

MPPos, in library 1 which were rich in Npcb were found to be cytotoxic with LD_{50} values < 40 µM and might not be suitable as transporter vehicles, but due to their toxicity to cancer cells they could serve as new anticancer agents. Their toxicity was significant higher to cancer cells compared to primary fibroblasts in two-dimensional cell culture tests. In addition, they efficiently inhibited the growth of three-dimensional tumor spheroids and displayed high uptake in the spheroids in a co-cultivation of endothelial cells, fibroblasts and spheroids in a microfluidic in vitro cancer model (vasQchip). For library 2 results for cytotoxicity assays were comparable to library 1, as highly lipophilic peptoids, rich in Npfb, induced cell death or inhibited cell growth. In contrast, low ratios of Npob in the peptoids revealed high cellular uptake, mitochondria penetration and no or only moderate cytotoxicity.

In the second part, libraries were screened for organ specificity in zebrafish larvae as an in vivo model, highly suitable for HTS of cell penetrating peptoids, as already shown by Franziska Rönicke [91]. According to their fluorescent phenotype zebrafish larvae could be classified in seven organ-specific categories, for library 1: apoptotic cells, eye/yolk sac, digestive system, kidney, caudal vein, neuromast cells and olfactory system. Structure-function evaluations demonstrated that side chain composition and hydrophilicity/hydrophobicity of peptoids was leading to specific accumulation of peptoids. Hydrophilic peptoids rich in Nlys displayed high affinity to neuromast cells, matching the results from literature, as 4-(4-Diethylaminostyryl)-1-methylpyridinium iodide, a hydrophilic and positive charged molecule was usually used for staining of neuromast hair cells [223-225]. Lipophilic peptoids, containing Nprg, Nphe and Npcb were accumulating in the olfactory system. Peptoids accumulating in the olfactory system might be suitable transporter for brain specific drug delivery, as the olfactory neurons and the olfactory nerve offers a possible way entering the brain by bypassing the BBB [191, 201, 202]. Hence, a peptoid, strongly accumulating in the olfactory system, was further analyzed by microinjection of peptoid solution in the cardinal vein of zebrafish embryos. After 24 h incubation of zebrafish, a strong peptoid signal was found in the brain tissue, indicating that P116 (Npcb-Nphe-Nprg-Nphe-RhodB) is a suitable candidate for brain delivery. As the BBB of zebrafish embryos is not completely developed in early stages, further tests in adult fish are needed in the future. For library 2 the suitability of Npob could be confirmed by analyses of peptoids in zebrafish embryos, showing accumulation of peptoids rich in Npob in the olfactory system. Comparison of peptoids in the olfactory system of library 1 and 2 revealed that the lipophilicity of the peptoids played a key role in their organ specificity. Even though, olfactory peptoids of library 1 and library 2 display different compositions of side chains, their calculated LogP values are comparable. Calculated LogP values of olfactory peptoids matched with LogP values for diazepam and progesterone, which are both known to cross the BBB and display high uptake in the brain [226].

Another interesting category are cyclic peptoids, which can be obtained after head-to-tail or side chain cyclization. In this work side chain cyclization, using a 1,3- dipolar cycloaddition, on solid support was found to be highly efficient for combinatorial libraries, synthesized in IRORI MiniKans. Diverse, fluorescent labeled cyclic peptoid libraries, giving enough yield for *in vitro* and *in vivo* analyze, could be synthesized. Analyses of cyclic peptoids for cellular uptake and intracellular localization revealed that cyclic peptoids are differing from their linear counterparts, showing mainly endosomal uptake for lipophilic, aromatic peptoids and low mitochondrial uptake for hydrophilic peptoids. Endocytotic uptake of cyclic peptoids, containing for example Nphe, Npcb or Npbf could be due to π-π-interactions between aromatic rings, leading to aggregation of peptoids, preventing direct penetration of membranes. Thus, aromatic and lipophilic cyclic peptoids also display lower toxicity compared to their linear counterparts, as endosomal enrichment was less toxic to cells compared to mitochondrial accumulation. In the literature it is often assumed, that cyclization of peptides or peptoids increased the cell permeability due to reduction of flexibility. However, Kwon and Kodadek et al. could disprove this hypothesis for peptides [227]. Visual analysis of microcopy data for cyclic peptides in comparison to their linear counterparts confirmed, that cyclization does not generally improve cell permeability. *In vivo* analysis exhibited further advantages of cyclic peptides, as it revealed less unspecific accumulations of cyclic peptides compared to linear peptides.

For most *in vitro* and *in vivo* investigations of peptoids, described above, fluorescent labeling is essential. Therefore, all peptoids were labeled with rhodamine B, which displays strong fluorescent signals with low photobleaching effects, moderate toxicity to cells. However, the molecular weight ratio of peptoid to dye is almost equal and an influence on the peptoid localization and uptake had to be considered. A small library containing interesting peptoid sequences of library 1 was coupled to fluorescein and a cyanine dye and intracellular localization and organ specificity in zebrafish larvae was compared to rhodamine B labeled peptides. *In vitro* analysis in HeLa cells displayed low influences of the dye concerning uptake and localization. Cyanine labeled peptoids displayed similar localizations in cells as the rhodamine B labeled peptoids, even though their lipophilicity is increased. Same results were also found for hydrophilic fluorescein labeled peptoid, only lipophilic fluorescein labeled peptoids were mostly located in the cytosol and not, as expected, in mitochondria. A possible explanation could be the hydrophilicity of fluorescein, decreasing the lipophilicity of the peptoid and the ability of direct cell membrane penetration. Unfortunately, organ-specific accumulation of cyanine and fluorescein labeled peptoids could not be confirmed in *in vivo* model. Due to the higher complexity of zebrafish larvae compared to 2D cell culture, increased lipophilicity of cyanine peptoids made the skin to a non-penetrating barrier. In contrast, fluorescein peptoids were mainly located in the digestive system. Hence, organ-specific localization could only be maintained by coupling to rhodamine B.

In the last part of this work fluorescent polymeric nanoparticles were investigated for drug delivery. In comparison to peptoids they had a large surface, giving the possibility for functionalization and increased drug to transporter ratio. They displayed uptake in the cytosol, an important feature for transport of drugs, like siRNA. In contrast to many other nanoparticles, investigated for drug delivery, their cytotoxic impacts are low with an LD_{50} of 150 µg/ml in HeLa cells, no haemolytic effect and moderate impacts on the development of zebrafish embryos. In comparison to P116, SCNPs were not found in the brain tissue after injection in the blood circulation, assuming that their size and hydrophilicity hinders them to cross the BBB. However, after injection in the brain ventricle accumulation in the brain tissue was detectable for several hours. As both, peptoids and nanoparticles, have their advantages and disadvantages it cannot be determined which DDS is more suitable in terms of drug delivery.

5. Material and Methods

5.1. Material

12-well plates	CELLSTAR®, Greiner
1-Hydroxybenzotriazolehydrate	Sigma-Aldrich
24-well plates	CELLSTAR®, Greiner
2-Butanol	Sigma-Aldrich
400 MHz NMR Spectrometer	Bruker
4-Chlorobenzylamine	Sigma-Aldrich
4-Fluorobenzylamine	Sigma-Aldrich
4-Hydroxybenzylamine	Sigma-Aldrich
96-well plate reader	SpectraMax ID3, Molecular Devices, USA
96-well plate reader	Ultra Microplate ELx808, BioTEK Instruments, INC
96-well plates	CELLSTAR®, Greiner
Acetonitril	Fisher Scientific
Agarose	Carl Roth
Agarose – low melting	VWR
Analytical balance	VWR
Ascorbic acid	Sigma-Aldrich
Benzylamine	Sigma-Aldrich
Boc-anhydride	Sigma-Aldrich
Bromoacetic acid	Sigma-Aldrich
Bromopropylamine hydrobromide	Sigma-Aldrich
Butylamine	Sigma-Aldrich
CDCl$_3$	Eurisotop
Cell culture flask	CELLSTAR®, Greiner
Cellincubator	Binder
Centrifuge	Eppendorf, VWR
Clean bench for cell culture	Clean Air
Collagen	Enzo Life Science
Confocal Microscope	Leica SPE
Confocal Microscope	Zeiss LSM800
Cu(I)iodide	Sigma-Aldrich
Diaminobutan	Sigma-Aldrich
Dichlormethan	Carl Roth
Diethylether	VWR
Dimethylformamide	VWR

Dimethylsulfoxide	Carl Roth
DIPEA	Sigma-Aldrich
DMEM	Gibco®
DPBS	Gibco®
Eppendorf tubes 1.5 ml	Eppendorf
Eppendorf tubes 2.0 ml	Eppendorf
Ethanol	VWR
Ethylacetate	VWR
FCS	Gibco®
FDA	Sigma-Aldrich
H_2DCFDA	Sigma-Aldrich
H_2O_2	Carl Roth
Hemocytometer	Brand
Heptylamine	Sigma-Aldrich
Hexylamine	Sigma-Aldrich
Hoechst 33342	Invitrogen
HPLC	GE Healthcare Life Science
HPLC	PURIFLASH, INTERCHIM
IBIDI 8-well plate for cell culture	IBIDI®
IBIDI 96-well plate for cell culture	IBIDI®
IRORI Mini+B2:B79Kans	IRORI®
IRORI USB-Stick	IRORI®
Lithium heparin vessels	Vacuette
MALDI TOF/TOF	Applied Biosystems/MDS SCIEX
Microinjector	Eppendorf
MitoTracker® Green	ThermoFisher
MitoTracker® Red	ThermoFisher
MTT	Promega
Multichannel pipette	Eppendorf
N,N'-Diisopropylcarbodiimide	Sigma-Aldrich
Na_2SO_4	Sigma-Aldrich
NovaSyn TGR resin	Sigma-Aldrich
Octanol	Sigma-Aldrich
Penicillin-Streptomycin	Gibco®
Petri dish	CELLSTAR®, Greiner
Piperidine	Carl Roth
Pipettes (0.5 µl-1000 µl)	Eppendorf
Propargylamine	Sigma-Aldrich

Propidiumiodide	Sigma-Aldrich
Pyridine	VWR
Rhodamine B	Sigma-Aldrich
Rink Amide AM resin LL	Novabiochem
Screening Microscope	Olympus Scan^R
Sodium hydroxide	Carl Roth
Sodiumazide	Sigma-Aldrich
SpeedVac	Eppendorf
Tetrahydrofuran	Carl Roth
Tricaine	Sigma-Aldrich
Trifluoroacetic acid	Carl Roth
Triton-X-100	Sigma-Aldrich
Trypan blue	Sigma-Aldrich
Trypsin-EDTA	Gibco®
Zebrafish incubator	Binder

5.2. Analytics

NMR

A 400 MHz NMR spectrometer (AVANCE 400, Bruker, Germany) was used. Chemical shifts are given in parts per million (δ/ppm) downfield from tetramethylsilane (TMS) and referenced to D_2O (4.80 ppm) as internal standard. The description of signals includes: s = singlet, bs = broad singlet.

Mass spectrometry

Peptoids were measured with A 4800 Plus MALDI TOF/TOF Analyzer (Applied Biosystems/MDS SCIEX) with a nitrogen laser. As matrix α-Cyano-4-hydroxycinnamic acid (CHCA, 10 mg/ml) in acetronitrile/water (1:1) with 0.1% TFA was used. 1.0 µl of a matrix/sample mixture (1:1) were applied onto a 384 spot Bruker MALDI target plate and allowed to dry. Molecular fragments were given as a mass-to-charge ratio [m/z].

High performance liquid chromatography (HPLC)

For the purification of peptoids two different preparative HPLC systems have been used:

1) An ÄKTApurifier 10 (GE Healthcare Life Science) with a preparative reversed phase-C18-coloumn was used at a flow rate of 3.5 ml/min with a linear gradient (solvent A= 95% H_2O, 5% ACN, 0.1% TFA and solvent B= 95% ACN, 5% H_2O, 0.1% TFA) from solvent

A to solvent B in 60 min. Separation of peptoids was detected at λ = 214 nm, λ = 480 nm (fluorescein labeling), λ = 560 nm (rhodamine B labeling) and λ = 650 nm (cyanine labeling).

2) A PURIFLASH® 4125 (INTERCHIM) with a preparative reversed phase-C18-coloumn was used at a flow rate of 10 ml/min with a linear gradient (solvent A= 100% H_2O, 0.1% TFA and solvent B= 100% ACN, 0.1% TFA) from solvent A to solvent B in 40 min. Separation of peptoids was detected at λ = 214 nm, λ = 480 nm (fluorescein labeling), λ = 560 nm (rhodamine B labeling) and λ = 650 nm (cyanine labeling).

5.3. Synthesis of submonomers

tert-butyl (4-aminobutyl)carbamate

The synthesis of *tert*-butyl (4-aminobutyl)carbamate was carried out as reported by Schröder *et al.* and optimized as described in the following [69, 207]:

62.3 g (70.0 mL, 707 mmol, 1.00 eq.) of 1,4-diaminobutane was dissolved in 400 ml THF. 23.1 g (106 mmol, 0.15 eq.) Boc-anhydride, dissolved in 220 ml of THF, was added dropwise over 4 h and the reaction mixture was stirred for 16 h. After concentration *in vacuo* and resuspension with 150 mL cold water, the suspension was filtered. The aqueous filtrate was extracted with ethyl acetate (4 x 50 ml). The combined organic layers were washed with water (2 x 25 mL), 40 mL brine, dried over Na_2SO_4 and then concentrated *in vacuo*. The product was obtained as a slightly yellow oil, yield 15.65 g (78%).

1-Azidopropylamine

The synthesis of Azidopropylamine was carried out as reported by Bandyopadhyay *et al.* for 1-Azidoethylamine and modified described in the following:

3 g (1 eq., 15 mmol) bromopropylamine hydrobromide were dissolved in 25 ml water. Subsequently, 3 g sodiumazide (3 eq., 50 mmol) were added and reaction was stirred for 8 h at 80 °C under reflux. Reaction was cooled and 1 g sodium hydroxide was added and reaction was stirred for further 10 min. Afterwards diethyl ether and water were added and organic layer was collected and dried over anhydrous Na_2SO_4. Solvent was evaporated giving a light-yellow liquid.

NMR: ^{1}H NMR (CDCl$_3$, 500 MHz) δ(ppm): 3.3 (2H), 2.7 (2H), 1.6 (2H)

5.4.   Synthesis peptoid libraries

Peptoid library 1

The synthesis of the cell penetrating peptoid library was carried out as reported by Kölmel et al. using permutation of four different side chains (N-2-prop-2-yn-1-ylglycine (Nprg), N-(p-chlorobenzyl)glycine (Npcb), N-4-aminobutylglycine (Nlys) and N-benzylglycine (Nphe) and is decribed in the following [91]:

Fmoc-protected Rink amide resin (AM resin LL 100-200 mesh, 0.61 mmol/g, 50 mg per IRORI-Kan) was swollen in DMF overnight. Afterwards the resin was washed 3 x 20 min with DMF. The following deprotection step was done 2 x 1 h with 20% piperidine in DMF. Between the two deprotection steps the Kans were washed 3 x 20 min with DMF. The acylation of the resin was done with 21.6 g bromoacetic acid (156 mmol, 20 eq.) and 24.1 ml diisopropylcarbodiimide (19.7 g, 156 mmol, 20 eq.) in 500 ml DMF for 2 h. After the reaction the Kans were washed 3 x 20 min with DMF. For the coupling with the different submonomers the Kans were incubated with 500 ml of a 1 M submonomer solution in DMF overnight. Before the next acylation the Kans were washed 3 x 20 min with DMF. Rhodamine B was coupled by overnight incubation with 67.7 g rhodamine B (78 mmol, 10 eq.), 24.1 ml diisopropylcarbodiimide (19.7 g, 156 mmol, 20 eq.) and 21.0 g (156 mmol, 20 eq.) 1-hydroxybenzotriazol in 500 ml DMF. After coupling the dye, the Kans were washed with DMF until the washing solution was clear. Kans were washed three times with DCM and the final cleavage was achieved by incubating the resin 2 h with 95% trifluoroacetic acid in DCM. All peptoids were dried in a SpeedVac and afterwards solved in water-ethanol (1:1). All peptoids were analyzed by MALDI-TOF MS and purified by HPLC (RP C18 column, buffer 1: 95% H_2O, 5% ACN, 0.1% TFA, buffer 2: 95% ACN, 5% H_2O, 0.1% TFA).

Peptoid library 2

CPPo library 2 was synthesized in collaboration with M.Sc. Bettina Fleck according to the protocol for library 1 using a permutation of three different side chains N-(p-fluorobenzyl)glycine (Npbf), N-benzylglycine (Nphe) and N-(p-hydroxybenzyl)glycine (Npob).

Peptoid library 3

Peptoids in library 3 were synthesized in the Molecular Foundry at Lawrence Berkeley National Laboratory, Berkeley, CA, USA. Five different side chains, tert-butyl (4-aminobutyl)carbamate, benzylamine, 4-chlorobenzylamine, butylamine and heptylamine, were used. All peptoids were synthesized on a fully automated synthesizer using 20 eq. bromoacetic acid and 24 eq. DIC for acylation (5 min) and 1 M submonomer solutions (20 min). Rhodamine B was coupled by overnight incubation with 5 eq. rhodamine B, 10 eq. DIC and 10 eq. HOBt in DMF. After coupling the dye resin was washed with DMF until the washing solution was clear. After three washing steps with DCM the final cleavage was achieved by incubating the resin 1 h with 95% trifluoroacetic acid in DCM.

Peptoid Library 4

Cyclic peptoids were synthesized as reported by Holub et al., with modifications for split-mix synthesis in IRORI MiniKans [137]. Eight different side chains (N-2-prop-2-yn-1-ylglycine (Nprg), N-(p-chlorobenzyl)glycine (Npcb), N-4-aminobutylglycine (Nlys), N-(p-fluorobenzyl)glycine (Npbf), N-(p-hydroxybenzyl)glycine (Npob), N-4-azidobutylglycine (Naz), N-hexylglycine (Nhex) and N-benzylglycine (Nphe) were used.

NovaSyn® TGR resin (0,23 mmol/g, 50 mg per IRORI-Kan) was swollen in DMF for 1 h. Afterwards the resin was washed 3 x 20 min with DMF. The acylation of the resin was done with 20 eq. bromoacetic acid and 20 eq. diisopropylcarbodiimide in DMF for 2 h. After the reaction the Kans were washed 3 x 20 min with DMF. For the coupling with the different submonomers the Kans were incubated with a 1 M submonomer solution in DMF overnight. All peptoids were coupled to Nprg in position 2 and Naz in position 5. Before the next acylation the Kans were washed 3 x 20 min with DMF. Cyclization of peptoids was done by incubating Kans overnight with 26 eq. Cu(I)iodide, 14 eq. ascorbic acide, 34 eq. DIPEA in 2-butanol/DMF/pyridine, 5:3:2 (0.2 ml/mg resin). Subsequently, Kans were washed 3 x 20 min with DMF, 3 x 20 min with DCM, 5 x 20 min with Pyridine/DMF, 1:1, with 0.2 g/10 ml ascorbic acide and again 3 x 20 min with DMF. Rhodamine B was coupled by overnight incubation with 20 eq. rhodamine B, 40 eq. diisopropylcarbodiimide and 40 eq. 1-hydroxybenzotriazol in DMF. After coupling the dye, the Kans were washed with DMF until the washing solution was clear. Kans were washed three times with DCM and the final cleavage was achieved by incubating the resin 2 h with 95% trifluoroacetic acid in dichloromethane. All peptoids were dried in a SpeedVac and afterwards solved in water-ethanol (1:1). All peptoids were analyzed by MALDI-TOF MS and purified by HPLC (RP C18 column, buffer 1: 95% H_2O, 5% ACN, 0.1% TFA, buffer 2: 95% ACN, 5% H_2O, 0.1% TFA).

Peptoids with cyanine dye and fluorescein

Cyanine dye (2-((1E,3Z,5E)-3-((2-carboxyethyl)thio)-5-(3-isopentyl-1,1-dimethyl-1,3-dihydro-2H-benzo[e]indol-2-ylidene)penta-1,3-dien-1-yl)-3-isopentyl-1,1-dimethyl-1H-benzo[e]indol-3-ium) was synthesized by Alexander Braun according to following protocol:

In a vial S2086 (500 mg, 0.640 mmol, 1.0 eq.) and mercaptropropionic acid (0.250 ml, 2.97 mmol, 4.6 eq.) were dissolved in 12.5 ml dry DMF and stirred for 48 h at 100 °C. The solvent was removed using reduced pressure. The left overs where washed with water and the water phase was extracted (3 × 100 ml) with ethyl acetate. The combined organic phases were washed with aq. NaCl-solution and dried over $MgSO_4$. The solvent was removed using reduced pressure. After purification by flash chromatography (starting with ethyl acetate/methanol 4:1 to methanol/ethyl acetate 1:1) 120 mg, (0.17 mmol, 26%) of a blue solid could be isolated as the desired product. – R_f = 0.083 (EA/MeOH 4:1). – ^{1}H-NMR (400 MHz, $CDCl_3$): δ (ppm) = 8.22-8.18 (m, 2 H, CH_{ar}), 8.11 (d, $^3J_{HH}$ = 8.4 Hz, 2 H, CH_{ar}), 7.95-7.92 (m, 4 H, CH_{ar}), 7.62 (t, $^3J_{HH}$ = 8.2 Hz, 2 H, CH_{ar}), 7.48 (t, $^3J_{HH}$ = 7.2 Hz, 2 H, CH_{ar}), 7.34 (d, $^3J_{HH}$ = 8.8 Hz, 2 H, CH), 7.00 (d, $^3J_{HH}$ = 13.7 Hz, 2 H, CH), 4.23 (m, 4 H, CH_2), 3.05 (t, $^3J_{HH}$ = 6.9 Hz, 2 H, CH_2), 2.62 (t, $^3J_{HH}$ = 6.9 Hz, 2 H, CH_2), 2.01 (s, 12 H, CH_3), 1.82-1.71 (m, H, CH/CH_2), 1.11 (d, $^3J_{HH}$ = 6.5 Hz, 12 H, CH_3).

Peptoids were synthesized as described for library 1 and 2 in IRORI MiniKans and subsequently coupling of dyes was done by overnight incubation with 5 eq. dye, 10 eq. diisopropylcarbodiimide and 10 eq. 1-hydroxybenzotriazol in DMF. After coupling the dyes, the Kans were washed with DMF until the washing solution was clear. Kans were washed three times with DCM and final cleavage was achieved by incubating the resin 2 h with 95% trifluoroacetic acid in dichloromethane. All peptoids were dried in a SpeedVac and afterwards solved in water-ethanol (1:1). All peptoids were analyzed by MALDI-TOF MS and purified by HPLC (RP C18 column, buffer 1: 95% H_2O, 5% ACN, 0.1% TFA, buffer 2: 95% ACN, 5% H_2O, 0.1% TFA).

5.5. Cell culture

Cancer cells, RAW cells and NHDF cells were incubated with DMEM medium supplemented with 10% FCS and 1% P/S at 37 °C and 5% CO_2 atmosphere. HUVEC cells were incubated with EGM-2 medium at 37 °C and 5% CO_2 atmosphere. For confocal microscopy cells were plated in 8-well IBIDI plates (1.5 x 10^4 cells per well in 200 µl DMEM). For automated screening microscopy cells were plated in 96-well IBIDI plates (1.5 x 10^4 cells per well in 200 µl DMEM).

Treatment with Peptoids

Cells were treated with 10 µM peptoids and incubated for 24 h at 37 °C and 5% CO_2 atmosphere. Afterwards 125 nM MitoTracker® Green was added and incubated for 30 min. Cells were washed 3 x with DPBS and finally DMEM medium with Hoechst 33342 (2 µg/ml) was added.

Confocal microscopy

Confocal microscopy was done with a Leica TCS-SPE microscope (DM2500) and a Zeiss LSM800 microscope. Experiments were done with different objectives: Leica TCS-SPE: ACS APO 10x/0.30 DRY, 20x/0.70 DRY UV and ACS APO 63x/1.30 OIL, Zeiss LSM800: Plan-Apochromat 20x/0.8 M27, Plan-Apochromat 63x/1.40 Oil DIC. Hoechst 33342 was excited with 405 nm (Em.: 417-468 nm). MitoTracker® Green at 488 nm (Em.: 499-552 nm) and MitoTracker® Red at 560 nm (Em.: 590-690 nm). Fluorescein labeled peptoids were excited at 488 nm (Em. 499-552 nm), rhodamine B labeled peptoids were excited at 561 nm (Em.: 593-696 nm) and cyanine labeled peptoids were excited at 633 nm (Em.: 650-750 nm).

Screening microscopy

Screening was done with an Olympus Scan^R IX81 microscope with a Hamamatsu C8484 camera and a 40x objective. For each well pictures were taken at 9 different positions, in each channel (Hoechst 33342, MitoTracker® Green and Peptoid).

LED Microscope

Growing of spheorids and toxicity of peptoids to zebrafish embryos was analyzed with a Leica DMIL LED microscope. Images were taken in brightfield using a HI PLAN 4x/0.10 DRY objective.

Cytotoxicity

The cytotoxicity of the different peptoids as well as SCNPs was determined with a CellTiter 96® Non-radioactive Cell Proliferation Assay (Promega), a modification of the 3-(4,5-dimethylthiazol-2-yl)-2,5-di-phenyltetrazolium bromide (MTT) assay. Due to intracellular reduction of tetrazolium salt to formazan in viable cells, viability can be measured, as formazan is detectable at 595 nm. Each well of a 96-well plate (Cstar 3596, 96 Well Cell Culture

Cluster, sterile) was seeded with 0.75×10^4 cells (HeLa, HepG2, MCF-7 and NHDF) in DMEM supplemented with 10% FCS and 1% P/S at 37 °C, 5% CO_2. For toxicity assays with HUVEC cells 96-well plates were pre-coated with 0.1 mg/ml collagen in DPBS and subsequently 1.0×10^5 cells/well were seeded in EGM-2 medium. After 24 h, medium was removed and cells were treated with medium (DMEM/EGM-2) containing the final concentrations. For each concentration, three wells were prepared and incubated for 72 h. As a control each plate contained three positive controls (cells treated with 5 µL of 20% triton) and three negative (untreated cells) control wells. After incubation 15 µl MTT solution was added and incubated for 4 h at 37 °C. Subsequently, reaction was stopped by adding 100 µl Solubilization/Stop solution to each well. After further incubation for 24 h the absorbance at 595 nm was recorded using a 96-well plate reader (SpectraMax ID3, Molecular Devices, USA). Data were averaged and the multiple determination of each concentration made it possible to calculate the standard deviation.

Analysis of cell death pathway

Rhodamine B labeled peptoids

1.5×10^4 HeLa cells were treated with peptoids in the respective LD_{50} concentration, which was determined by cytotoxicity assays, for 24 h at 37 °C and 5% CO_2 atmosphere. Cells were washed 3 x with DPBS and DMEM medium with Hoechst 33342 (2 µg/ml) was added. Finally, trypan blue 50 µl, 0.4%) was added to the medium and cells were analyzed by confocal microscopy.

Peptoids without rhodamine B

1.5×10^4 HeLa cells were treated with peptoids in the respective LD_{50} concentration, which was determined by cytotoxicity assays, for 24 h at 37 °C and 5% CO_2 atmosphere. Subsequently, medium was removed and medium containing 8 µl/5 ml medium FDA (5 mg/ml) and 50 µl/5 ml medium PI (2 mg/ml) was added and cells were incubated for 5 min at room temperature. Cells were washed 3 x with DPBS and DMEM medium with Hoechst 33342 (2 µg/ml) was added and staining of cells was investigated by confocal microscopy.

Analysis of reactive oxygen species

1.5×10^4 HeLa cells were treated with 30 µM peptoid solution for 4 h at 37 °C and 5% CO_2 atmosphere. 1 h before end of incubation time 5 µM H_2DCFDA was added. Afterwards cells were washed 3 x with DPBS and DMEM medium with Hoechst 33342 (2 µg/ml) was added. Furthermore, a positive control (H_2O_2) and negative control (untreated cells) was done.

Spheroids

A 96-well plate (Cstar 3596, 96 Well Cell Culture Cluster, sterile) was filled with 50 µl 1.5% agarose per well. After hardening of agarose 50 µl SK-MEL 28 cells in DMEM (4000 cells per well) were added. Cells were incubated for 24 h at 37 °C and 5% CO_2 atmosphere. Afterwards, growing of spheroids was enhanced by adding 50 µl DMEM. After 72 h peptoid solution was added to the medium. For analysis of cell penetration 10 µM peptoid was added and incubated overnight. Subsequently, spheroids were washed with DPBS and analyzed by confocal microscopy. For growing curves for each peptoid six spheroids were analyzed. Respectively, three spheroids were incubated with 20 µM peptoid and three spheroids were incubated with 40 µM peptoid. Growing of cells was analyzed by microscopy after 1, 2, 3 and 4 days of incubation. Furthermore, a set of untreated spheroids was analyzed. To determine the size of the spheroids diameters of each spheroid were defined with the ImageJ software. Data were averaged and standard deviation were calculated.

VasQchips

VasQchips were fabricated as described by Kappings et al. [167, 168] and experiments were done in collaboration with Dr. Eva Zittel. The microchannel of the *vasQchips* was coated with 0.9 mg/ml fibronectin and the lower compartment was coated with 0.6 mg/ml collagen type I solution (1 h). Subsequently chips were filled with EGM-2 medium and 5×10^5 HUVEC cells were added to the microchannel. To achieve a uniform cell attachment, cells were rotated (0.25 rpm, 1 h). The system was incubated for one week under microfluidic, with a flow rate from 100 µl/min to 450 µl/min, using a pump system (IBIDI). The medium was exchanged every second day. Afterwards the lower compartment was filled with a fibringel according to Nakatsu et al. [228]: (4 mg/ml fibrinogen solution, 0.15 U/ml Aprotinin, 22.5 µg/ml Collagen, 0.625 U/ml Thrombin. Furthermore, NHDF cells (1×10^6) and SK-MEL 28 spheroids (5-10) were added. Chips were incubated overnight and afterwards cell nuclei were stained with Hoechst

33342 and cells were analyzed by confocal microscopy (Leica TCS-SPE, ACS APO 10xDRY objective). Subsequently, 10 μM peptoid solution was added to the medium in the microfluidic system and uptake of peptoids in cells was analyzed after 24, 48 and 96 h.

5.6.   Automated image processing

Automated image processing and data analysis was done by Prof. Dr. Ralf Mikut and Dr. Markus Reischl (Institute for Automation and Applied Informatics, KIT) according to following protocol:

Image processing was done using MATLAB 2016b. The pixel-values of images in the DAPI-channel were normalized to [0, 1]. An OTSU-thresholding then delivered an estimation of the nucleus. A subsequent dilation (disc, radius 9 pixels) and subtraction of the threshold image delivered a ring around the nucleus.

To quantify fluorescent-activity, GFP resp. RFP images were background-corrected (each pixel is subtracted the minimum-value within a 40x40px region). The image was then subtracted the 25th percentile of brightness-values, negative values are set to 0. A convolution (5x5pixels) suppresses noise. All pixels exceeding afterwards 10% of the maximum brightness are counted as fluorescent (e.g. GFP positive).

An image was then described by four features:

- Rfp_org: Mean value of RFP-images within all rings around the nucleus,
- Gfp_org: Mean value of GFP-images within all rings around the nucleus,
- Rfp_on_Gfp: Quantity of fluorescent activity: Portion of processed GFP-pixels that also show RFP-signal within the rings around the nucleus [0 … 1] and
- Cellarea: Portion of space settled by nuclei within an image [0...100].

Data analysis

The data from image processing were analyzed with the Peptide Extension of the MATLAB Toolbox Gait-CAD [229]. The raw dataset contains 6777 single position measures for 234 peptoids in 753 wells. Position measurements with unusual large cell area values and unusual high GFP values were automatically marked as artifact and excluded using a Fuzzy C Means clustering. As a result, 6125 artifact free positions measurements for 233 peptoids in 711 wells were used for the following steps. Mean values were computed for all 711 wells and all features from image processing. These mean values were aggregated to 233 mean values of all wells containing the same peptoid. These results were visualized as boxplots or two-dimensional histograms depending on the proportion of all submonomers in a peptoid sequence or statistically analyzed to find relevant differences.

5.7. Molecular Dynamics Simulations

Simulations were performed by Daniel Holub (Institute of Physical Chemistry, KIT) according to following protocol:

All simulations were performed with GROMACS 5.0.4 and tools of the program were used to create the system [230, 231]. The starting structures for the simulations of the rhodamine B labeled peptoid simulations were created with the program AVOGADRO. The peptoid was placed in the center of a 5.5 x 5.5 x 7 nm large box and solvated with water using the TIP3P model. This box was extended along the z-axis to double the size (5.5 x 5.5 x 14 nm) and resulting space was filled up with octanol molecules. The final system contains 7966 solvent molecules, 633 octanol molecules and one peptoid molecule. The general Amber force field (GAFF) was applied to parametrize octanol, the rhodamine B labeled peptoid with the *xLeap* module of AmberTools [232-234]. The atomic charges were calculated with restrained fitting of the electrostatic potential (RESP) obtained at the HF/6-31G* level of theory with GAUSSIAN09 [235-239]. To generate a neutral system respectively sodium or chlorin ions were added. The system was equilibrated by means of a protocol consisting of a series of energy minimization, NVT and NPT simulations. The eventual, production simulation used the Nosé-Hoover thermostat and the Parrinello-Rahman barostat to maintain a temperature of 300 K and a pressure of 1 bar, respectively [240, 241]. The simulations employed a leap-frog integrator with a time step of 2 fs and were extended to 300 ns.

5.8. Zebrafish treatments

Zebrafish husbandry

Adult zebrafish (Danio rerio, Casper line (mitfa$^{w2/w2}$; roy$^{a9/a9}$)) were bred and maintained according to standard methods (Westerfield 2007) in a 14 h on/10 h off light cycle. Fish were crossed pair-wise and eggs were collected within 3 h of laying and incubated in petri dishes filled with E3-medium (5 mM NaCl, 0,17 mM KCl, 0,33 mM CaCl$_2$, 0.33 mM MgSO$_4$) and 1 mg/ml of methylene blue. E3-medium was daily renewed until the beginning of the experiment.

Zebrafish treatment with peptoids

For screening of peptoid libraries 96 hpf larvae were incubated in 50 µM peptoid solution in E3-medium solution for 2 h. For each peptoid 8 larvae were treated with the same peptoid. After incubation, larvae were washed with E3 medium for three times, anesthetized with 0.02% tricaine and transferred to 8-well IBIDI plates for investigation by confocal microscopy (Leica TCS-SPE microscope) or 96-well IBIDI plates for high-throughput screenings (Olympus Scan^R IX81 microscope with a Hamamatsu C8484 camera). Furthermore, each plate contained 8 control zebrafish, which were incubated only with E3-medium.

Zebrafish injection: Cardinal vein

Zebrafish were dechorionated 48 hpf and anesthetized with 0.02% tricaine. 15 mg/ml SCNP / 1mM peptoid solution in E3-medium was injected in the blood circulation by microinjection in the cardinal vein. Afterwards treated zebrafish were transferred in E3-medium and incubated for 24 h, allowing them to recover, at 27 °C. Subsequently the embryos were anesthetized with 0.02% tricaine and analyzed by confocal microscopy.

Zebrafish injection: Brain ventricle

Zebrafish were dechorionated 48 hpf and anesthetized with 0.02% tricaine. 15 mg/ml SCNP solution in E3-medium was injected in the brain ventricle by microinjection. Afterwards treated zebrafish were transferred in E3-medium and incubated at 27 °C. After 2 h and 24 h larvae were anesthetized with 0.02% tricaine and embedded in 1% low-melting agarose bevor analyzing by confocal microscopy.

Fish embryo toxicity test

Fertilized eggs, 4 hpf, were sorted in a 96-well plate (one embryo per well) and were incubated in 200 µl E3-medium with increasing concentrations of SCNPs (7.5, 15, 30, 75 and 100 µg/ml) Embryos were incubated at 27 °C and analyzed each day (24 hpf, 48 hpf, 72 hpf, 96 hpf) by microscopy using a Leica DMIL LED microscope. Hatching rate, mortality, degradation of movements and deformation rates were examined.

Fish embryo toxicity test, dechorionated fish

Zebrafish were dechorionated 24 hpf and sorted in a 96-well plate (one embryo per well) and were incubated in 200 µl E3-medium with increasing concentrations of peptoids (5 µM,

10 µM, 25 µM, 50 µM). Embryos were incubated at 27 °C and analyzed each day (48 hpf, 72 hpf, 96 hpf, 120 hpf) by microscopy. Mortality, degradation of movements and deformation rates were examined.

5.9. Octanol-water partition coefficient

Molar attenuation coefficient of Rhodamine B in water and octanol at l=550 nm was determined by diluting Rhodamine B in octanol and water to a final concentration of 20 µM. The solution was measured in a 96-well-plate (Cstar 3596, 96 Well Cell Culture Cluster, sterile), 200 µL per well, by using a 96-well plate reader (Ultra Microplate Reader ELx808, BioTEK Instruments, INC). Peptoids were diluted to a final concentration of 160 µM in 250 µL water. Afterwards 250 µL octanol was added and each mix was vortexed for 2 min and centrifuged for 3 min with a centripetal force of 3000 g to separate the octanol from the aqueous phase. Phases were separated and the absorbance at l=550 nm of 200 µL octanol phase and water phase was measured in a 96-well-plate (Cstar 3596, 96 Well Cell Culture Cluster, sterile) by using a 96-well plate reader (Ultra Microplate Reader ELx808, BioTEK Instruments, INC). Each absorbance of peptoid was measured at least three times. Data were averaged and standard deviation was calculated.

5.10. Haemolysis

Hemolytic activity of SCNPs was examined using blood from rats collected in lithium heparin vessels. Blood was centrifuged at 3000 g for 10 min. Afterward supernatant was discarded and red blood cells were washed with DPBS until a clear supernatant was observed. Erythrocytes were diluted in DPBS to 4% (v/v) and 50 µl were mixed with 50 µl SCNP solution in DPBS in 1.5 ml Eppendorf tubes. As a positive control 50 µl 1% Triton X-100 and as negative control DPBS was used. Cells were incubated at 37 °C while shaking for 1 h. Afterwards cells were centrifuged at 3000 g for 10 min and the supernatant was transferred in a 96- well plate. Absorbance was measured at 540 nm using a 96-well plate reader (Ultra Microplate Reader ELx808, BioTEK Instruments, INC). The percentage lysis was calculated relative to 0% lysis with DPBS and 100% lysis with Triton X-100.

6. Abbreviations

µl	Microliter
µM	Micromolar
µm	Micrometer
λ	Wavelength
°C	Celsius
2D	Two dimensional
3D	Three dimensional
A	Alanine
ACN	Acetonitrile
BBB	Blood-brain-barrier
Boc	Tert-butoxycarbonyl
C	Cysteine
CSF	Cerebrospinal fluid
CO_2	Carbondioxide
CNS	Central nervous system
CPP	Cell-penetrating peptide
CPPo	Cell penetrating peptoide
CuAAc	Copper(I)-catalyzed azide alkyne cycloaddition
Cy	Cyanine dye
DCM	Dichlormethane
DCF	2',7'-Dichlorfluorescein
DDS	Drug delivery system
DIC	Diisopropylcarbodiimide
DIPEA	Diisopropylethylamin
Dox	Doxorubicin
DMA	Droplet microarrays
DMEM	Dulbecco's Modified Eagle's Medium
DMF	Dimethylformamide
DMSO	Dimethylsulfoxid
DNA	Deoxyribonucleic acid
DPBS	Dulbecco's Phosphate-Buffered Saline

E	Glutamate
EDTA	Ethylenediaminetetraacetic acid
EPR	Enhanced permeability and retention
Em.	Emission
Eq.	Equivalent
EtO2	Diethylether
EtOH	Ethanol
Ex.	Excitation
F	Phenylalanine
Fmoc	Fluorenylmethyloxycarbonyl protecting group
FET	Fish embryo toxicity
FDA	Fluorescein diacetate
Fluo	Fluorescein
G	Glycine
g	Gramm
GFP	Green fluorescent protein
h	Hour
H$_2$O$_2$	Hydrogen peroxide
H$_2$DCFDA	2',7'-Dichlordihydrofluorescein-diacetat
HeLa	Cervical cancer cell line
HepG2	Liver cancer cell line
HIV	Human immunodeficiency virus
HOBt	Hydroxybenzotriazole
hpf	Hours post fertilization
HPLC	High-performance liquid chromatography
HUVEC	Human umbilical vein endothelial cells
HTS	High-throughput screening
I	Isoleucine
ITG	Institute of Toxicology and Genetics
KIT	Karlsruher Institute of Technology
L	Leucine
LC/MS	Liquid chromatography–mass spectrometry

LD$_{50}$	Lethal dose of 50%
M	Methionine
MALDI	Matrix-assisted laser desorption/ionization
MeOH	Methanol
MCF-7	Breast cancer cell line
Min	Minute
ml	Milliliter
mm	millimeter
mM	Millimolar
mmol	Millimol
MTT	3-(4,5-dimethylthiazol-2-yl)-2,5-diphenyltetrazolium bromide
MS	Mass spectrometry
N	Asparagine
*N*lys	Diminobutan peptoid side chain
*N*prg	Propargylamine peptoid side chain
*N*phe	Benzylamine peptoid side chain
*N*pcb	p-Chlorobenzylamine peptoid side chain
*N*pbf	p-Fluorobenzylamine peptoid side chain
*N*pob	p-Hydroxybenzylamine peptoid side chain
*N*but	Butylamine peptoid side chain
*N*4az	4-Azidopropargylamine peptoid side chain
Nhep	Heptylamine peptoid side chain
Na$_2$SO$_4$	Natriumsulfate
NHDF	Normal Human Dermal Fibroblasts
nm	Nanometer
nM	Nanomolar
OBOC	One bead one compound
ORN	Olfactory receptor neurons
P	Proline
PAA	Poly(acrylic acid)
P/S	Penicillin *and* streptomycin
Q	Glutamine

R	Arginine
rf	Radiofrequence
RFP	Red fluorescent protein (Red signal)
RhodB	Rhodamine B
ROI	Region of interest
ROS	Reactive oxygen species
RNA	Ribonucleic acid
SCNP	Single chain nanoparticles
S	Serine
RT	Room temperature
TADF	Thermally Activated Delayed Fluorescence
tat	Trans-Activator of Transcription
TFA	Trifluoroacetic acid
TOF	Time of flight
UV	Ultraviolet
V	Valine
Vit. C	Vitamin C
W	Tryptophan
Y	Tyrosine
zfGUI	Zebrafish graphical user interface
MPP	Mitochondria penetrating peptide
MPPo	Mitochondria penetrating peptoid

7. References

1.   Torchilin, V.P., *Drug targeting*. European Journal of Pharmaceutical Sciences, 2000. **11**: p. S81-S91.
2.   Classen, D.C., et al., *Adverse drug events in hospitalized patients: Excess length of stay, extra costs, and attributable mortality*. JAMA, 1997. **277**(4): p. 301-306.
3.   Langer, R., *Drug delivery and targeting*. Nature, 1998. **392**(6679 Suppl): p. 5-10.
4.   Scherrmann, J.M., *Drug delivery to brain via the blood-brain barrier*. Vascul Pharmacol, 2002. **38**(6): p. 349-54.
5.   Pardridge, W.M., *Drug delivery to the brain*. J Cereb Blood Flow Metab, 1997. **17**(7): p. 713-31.
6.   Pardridge, W.M., *Blood-brain barrier delivery*. Drug Discov Today, 2007. **12**(1-2): p. 54-61.
7.   Patel, J.P. and B.N. Frey, *Disruption in the Blood-Brain Barrier: The Missing Link between Brain and Body Inflammation in Bipolar Disorder?* Neural plasticity, 2015. **2015**: p. 708306-708306.
8.   Brightman, M.W., *Morphology of blood-brain interfaces*. Experimental Eye Research, 1977. **25**: p. 1-25.
9.   Janzer, R.C. and M.C. Raff, *Astrocytes induce blood–brain barrier properties in endothelial cells*. Nature, 1987. **325**: p. 253.
10.   Abbott, N.J., et al., *Structure and function of the blood-brain barrier*. Neurobiol Dis, 2010. **37**(1): p. 13-25.
11.   Katzung, B.G., S.B. Masters, and A.J. Trevor, *Basic and Clinical Pharmacology (LANGE Basic Science)*. 2012: McGraw-Hill Education.
12.   Tamargo, J., J.-Y. Le Heuzey, and P. Mabo, *Narrow therapeutic index drugs: a clinical pharmacological consideration to flecainide*. European Journal of Clinical Pharmacology, 2015. **71**(5): p. 549-567.
13.   Maeda, H., *Tumor-selective delivery of macromolecular drugs via the EPR effect: background and future prospects*. Bioconjug Chem, 2010. **21**(5): p. 797-802.
14.   Mishra, N., et al., *Targeted Drug Delivery: A Review*. Vol. 6. 2016.
15.   Batist, G., et al., *Reduced Cardiotoxicity and Preserved Antitumor Efficacy of Liposome-Encapsulated Doxorubicin and Cyclophosphamide Compared With Conventional Doxorubicin and Cyclophosphamide in a Randomized, Multicenter Trial of Metastatic Breast Cancer*. Journal of Clinical Oncology, 2001. **19**(5): p. 1444-1454.
16.   Chowdhary, R.K., I. Shariff, and D. Dolphin, *Drug release characteristics of lipid based benzoporphyrin derivative*. J Pharm Pharm Sci, 2003. **6**(1): p. 13-9.
17.   Green, M. and P.M. Loewenstein, *Autonomous functional domains of chemically synthesized human immunodeficiency virus tat trans-activator protein*. Cell, 1988. **55**(6): p. 1179-88.
18.   Frankel, A.D. and C.O. Pabo, *Cellular uptake of the tat protein from human immunodeficiency virus*. Cell, 1988. **55**(6): p. 1189-1193.
19.   Sternson, L.A., *Obstacles to polypeptide delivery*. Ann N Y Acad Sci, 1987. **507**: p. 19-21.
20.   Joliot, A., et al., *Antennapedia homeobox peptide regulates neural morphogenesis*. Proceedings of the National Academy of Sciences of the United States of America, 1991. **88**(5): p. 1864-1868.
21.   Derossi, D., et al., *The third helix of the Antennapedia homeodomain translocates through biological membranes*. J Biol Chem, 1994. **269**(14): p. 10444-50.
22.   Pooga, M., et al., *Cell penetration by transportan*. Faseb j, 1998. **12**(1): p. 67-77.
23.   Murphy, A.L. and S.J. Murphy, *Catch VP22: the hitch-hiker's ride to gene therapy?* Gene Ther, 1999. **6**(1): p. 4-5.
24.   Zaro, J.L. and W.-C. Shen, *Cationic and amphipathic cell-penetrating peptides (CPPs): Their structures and in vivo studies in drug delivery*. Frontiers of Chemical Science and Engineering, 2015. **9**(4): p. 407-427.
25.   Morris, M.C., et al., *Cell-penetrating peptides: from molecular mechanisms to therapeutics*. Biol Cell, 2008. **100**(4): p. 201-17.

26. Oehlke, J., et al., *Cellular uptake of an α-helical amphipathic model peptide with the potential to deliver polar compounds into the cell interior non-endocytically.* Biochimica et Biophysica Acta (BBA) - Biomembranes, 1998. **1414**(1): p. 127-139.

27. Rhee, M. and P. Davis, *Mechanism of uptake of C105Y, a novel cell-penetrating peptide.* J Biol Chem, 2006. **281**(2): p. 1233-40.

28. Guidotti, G., L. Brambilla, and D. Rossi, *Cell-Penetrating Peptides: From Basic Research to Clinics.* Trends in Pharmacological Sciences, 2017. **38**(4): p. 406-424.

29. Sugita, T., et al., *Comparative study on transduction and toxicity of protein transduction domains.* Br J Pharmacol, 2008. **153**(6): p. 1143-52.

30. Cardozo, A.K., et al., *Cell-permeable peptides induce dose- and length-dependent cytotoxic effects.* Biochim Biophys Acta, 2007. **1768**(9): p. 2222-34.

31. Toro, A., et al., *Intracellular delivery of purine nucleoside phosphorylase (PNP) fused to protein transduction domain corrects PNP deficiency in vitro.* Cell Immunol, 2006. **240**(2): p. 107-15.

32. Mueller, J., et al., *Comparison of Cellular Uptake Using 22 CPPs in 4 Different Cell Lines.* Bioconjugate Chemistry, 2008. **19**(12): p. 2363-2374.

33. Richard, J.P., et al., *Cell-penetrating peptides. A reevaluation of the mechanism of cellular uptake.* J Biol Chem, 2003. **278**(1): p. 585-90.

34. Vives, E., P. Brodin, and B. Lebleu, *A truncated HIV-1 Tat protein basic domain rapidly translocates through the plasma membrane and accumulates in the cell nucleus.* J Biol Chem, 1997. **272**(25): p. 16010-7.

35. Lundberg, M., S. Wikström, and M. Johansson, *Cell surface adherence and endocytosis of protein transduction domains.* Molecular Therapy, 2003. **8**(1): p. 143-150.

36. Palm-Apergi, C., P. Lönn, and S.F. Dowdy, *Do Cell-Penetrating Peptides Actually "Penetrate" Cellular Membranes?* Molecular Therapy, 2012. **20**(4): p. 695-697.

37. Duchardt, F., et al., *A comprehensive model for the cellular uptake of cationic cell-penetrating peptides.* Traffic, 2007. **8**(7): p. 848-66.

38. Conner, S.D. and S.L. Schmid, *Regulated portals of entry into the cell.* Nature, 2003. **422**: p. 37.

39. Aderem, A. and D.M. Underhill, *Mechanisms of phagocytosis in macrophages.* Annu Rev Immunol, 1999. **17**: p. 593-623.

40. Järver, P., I. Mäger, and Ü. Langel, *In vivo biodistribution and efficacy of peptide mediated delivery.* Trends in Pharmacological Sciences, 2010. **31**(11): p. 528-535.

41. Pouny, Y., et al., *Interaction of antimicrobial dermaseptin and its fluorescently labeled analogs with phospholipid membranes.* Biochemistry, 1992. **31**(49): p. 12416-12423.

42. Alves, I.D., et al., *Membrane interaction and perturbation mechanisms induced by two cationic cell penetrating peptides with distinct charge distribution.* Biochimica et Biophysica Acta (BBA) - General Subjects, 2008. **1780**(7): p. 948-959.

43. Cardoso, A.M.S., et al., *S4(13)-PV cell-penetrating peptide induces physical and morphological changes in membrane-mimetic lipid systems and cell membranes: Implications for cell internalization.* Biochimica et Biophysica Acta (BBA) - Biomembranes, 2012. **1818**(3): p. 877-888.

44. Koren, E. and V.P. Torchilin, *Cell-penetrating peptides: breaking through to the other side.* Trends in Molecular Medicine, 2012. **18**(7): p. 385-393.

45. Herce, H.D., et al., *Visualization and targeted disruption of protein interactions in living cells.* Nature Communications, 2013. **4**: p. 2660.

46. Virès, E., et al., *Structure-activity relationship study of the plasma membrane translocating potential of a short peptide from HIV-1 Tat protein.* Letters in Peptide Science, 1997. **4**(4): p. 429-436.

47. Kristensen, M., D. Birch, and H. Morck Nielsen, *Applications and Challenges for Use of Cell-Penetrating Peptides as Delivery Vectors for Peptide and Protein Cargos.* Int J Mol Sci, 2016. **17**(2).

48. Juliano, R., et al., *Mechanisms and strategies for effective delivery of antisense and siRNA oligonucleotides.* Nucleic Acids Research, 2008. **36**(12): p. 4158-4171.

49. Nagahara, H., et al., *Transduction of full-length TAT fusion proteins into mammalian cells: TAT-p27Kip1 induces cell migration.* Nature Medicine, 1998. **4**: p. 1449.

50. Morris, M.C., et al., *A peptide carrier for the delivery of biologically active proteins into mammalian cells.* Nature Biotechnology, 2001. **19**: p. 1173.

51. Jaroslaw Ruczynski, A.U.P.M.W.A.U.M.K.-W.A.U.P.M.A.U.K.S.-K.A.U.P.R., *Cell-penetrating peptides as a promising tool for delivery of various molecules into the cells.* Cell-penetrating peptides as a promising tool for delivery of various molecules into the cells, 2014. **52**(4): p. 257-269-257-269.

52. Shen, Y., et al., *A novel cell-penetrating peptide to facilitate intercellular transport of fused proteins.* J Control Release, 2014. **188**: p. 44-52.

53. Jo, J., et al., *Cell-penetrating peptide (CPP)-conjugated proteins is an efficient tool for manipulation of human mesenchymal stromal cells.* Scientific reports, 2014. **4**: p. 4378-4378.

54. Cao, G., et al., *In Vivo Delivery of a Bcl-xL Fusion Protein Containing the TAT Protein Transduction Domain Protects against Ischemic Brain Injury and Neuronal Apoptosis.* J Neurosci, 2002. **22**(13): p. 5423-31.

55. Schwarze, S.R., et al., *In vivo protein transduction: delivery of a biologically active protein into the mouse.* Science, 1999. **285**(5433): p. 1569-72.

56. Rousselle, C., et al., *New advances in the transport of doxorubicin through the blood-brain barrier by a peptide vector-mediated strategy.* Mol Pharmacol, 2000. **57**(4): p. 679-86.

57. van den Berg, A. and S.F. Dowdy, *Protein transduction domain delivery of therapeutic macromolecules.* Current Opinion in Biotechnology, 2011. **22**(6): p. 888-893.

58. Kerkis, A., et al., *Properties of cell penetrating peptides (CPPs).* IUBMB Life, 2006. **58**(1): p. 7-13.

59. Anderson, D.C., et al., *Tumor cell retention of antibody Fab fragments is enhanced by an attached HIV TAT protein-derived peptide.* Biochem Biophys Res Commun, 1993. **194**(2): p. 876-84.

60. Jain, M., et al., *Penetratin improves tumor retention of single-chain antibodies: a novel step toward optimization of radioimmunotherapy of solid tumors.* Cancer Res, 2005. **65**(17): p. 7840-6.

61. Palm, C., et al., *Peptide degradation is a critical determinant for cell-penetrating peptide uptake.* Biochimica et Biophysica Acta (BBA) - Biomembranes, 2007. **1768**(7): p. 1769-1776.

62. Vagner, J., H. Qu, and V.J. Hruby, *Peptidomimetics, a synthetic tool of drug discovery.* Curr Opin Chem Biol, 2008. **12**(3): p. 292-6.

63. Holm, T., et al., *Retro-inversion of certain cell-penetrating peptides causes severe cellular toxicity.* Biochimica et Biophysica Acta (BBA) - Biomembranes, 2011. **1808**(6): p. 1544-1551.

64. Horne, W.S., et al., *Structural and biological mimicry of protein surface recognition by α/β-peptide foldamers.* Proceedings of the National Academy of Sciences, 2009.

65. Spatola, A.F., *Peptide Backbone Modifications: a Structure-Activity Analysis of Peptides Containing Amide Bond Surrogates. Conformational Constraints, and Related Backbone Replacements.,* In Chemistry and Biochemistry of Amino Acids, Peptides, and Proteins, Volume 7 edn. ch, 1983. **5**: p. 267-357.

66. Avan, I., C.D. Hall, and A.R. Katritzky, *Peptidomimetics via modifications of amino acids and peptide bonds.* Chemical Society Reviews, 2014. **43**(10): p. 3575-3594.

67. Miller, S.M., et al., *Comparison of the proteolytic susceptibilities of homologous L-amino acid, D-amino acid, and N-substituted glycine peptide and peptoid oligomers.* Drug Development Research, 1995. **35**(1): p. 20-32.

68. Wang, Y., et al., *Absorption and disposition of a tripeptoid and a tetrapeptide in the rat.* Biopharm Drug Dispos, 1999. **20**(2): p. 69-75.

69. Schröder, T., et al., *Solid-Phase Synthesis, Bioconjugation, and Toxicology of Novel Cationic Oligopeptoids for Cellular Drug Delivery.* Bioconjugate Chemistry, 2007. **18**(2): p. 342-354.

70. Seo, J., B.C. Lee, and R.N. Zuckermann, *2.3 Peptoids: Synthesis, Characterization, and Nanostructures.* 2017.

71. Sanborn, T.J., et al., *Extreme stability of helices formed by water-soluble poly-N-substituted glycines (polypeptoids) with α-chiral side chains.* Biopolymers, 2002. **63**(1): p. 12-20.

72. Kapoor, R., et al., *Antimicrobial Peptoids Are Effective against Pseudomonas aeruginosa Biofilms.* Antimicrobial Agents and Chemotherapy, 2011. **55**(6): p. 3054-3057.

73. Lyczak, J.B., C.L. Cannon, and G.B. Pier, *Establishment of Pseudomonas aeruginosa infection: lessons from a versatile opportunist.* Microbes Infect, 2000. **2**(9): p. 1051-60.

74. Mojsoska, B., R.N. Zuckermann, and H. Jenssen, *Structure-Activity Relationship Study of Novel Peptoids That Mimic the Structure of Antimicrobial Peptides.* Antimicrobial Agents and Chemotherapy, 2015. **59**(7): p. 4112-4120.

75. Eggimann, G.A., et al., *Investigating the Anti-leishmanial Effects of Linear Peptoids.* ChemMedChem, 2015. **10**(2): p. 233-237.

76. Ryge, T.S., N. Frimodt-Møller, and P.R. Hansen, *Antimicrobial Activities of Twenty Lysine-Peptoid Hybrids against Clinically Relevant Bacteria and Fungi.* Chemotherapy, 2008. **54**(2): p. 152-156.

77. Lee, J., et al., *Prostate tumor specific peptide–peptoid hybrid prodrugs.* Bioorganic & Medicinal Chemistry Letters, 2015. **25**(14): p. 2849-2852.

78. Goodson, B., et al., *Characterization of Novel Antimicrobial Peptoids.* Antimicrobial Agents and Chemotherapy, 1999. **43**(6): p. 1429-1434.

79. Huang, W., et al., *Learning from Host-Defense Peptides: Cationic, Amphipathic Peptoids with Potent Anticancer Activity.* PLoS ONE, 2014. **9**(2): p. e90397.

80. Hooks, J.C., J.P. Matharage, and D.G. Udugamasooriya, *Development of homomultimers and heteromultimers of lung cancer-specific peptoids.* Biopolymers, 2011. **96**(5): p. 567-77.

81. Zuckermann, R. and T. Kodadek, *Peptoids as Potential Therapeutics.* Vol. 11. 2009. 299-307.

82. Gibbons, J.A., et al., *Pharmacologic characterization of CHIR 2279, an N-substituted glycine peptoid with high-affinity binding for alpha 1-adrenoceptors.* J Pharmacol Exp Ther, 1996. **277**(2): p. 885-99.

83. Gante, J., *Peptidomimetics—Tailored Enzyme Inhibitors.* Angewandte Chemie International Edition in English, 1994. **33**(17): p. 1699-1720.

84. Hamy, F., et al., *An inhibitor of the Tat/TAR RNA interaction that effectively suppresses HIV-1 replication.* Proceedings of the National Academy of Sciences, 1997. **94**(8): p. 3548-3553.

85. Mannige, R.V., et al., *Peptoid nanosheets exhibit a new secondary-structure motif.* Nature, 2015. **526**: p. 415.

86. Robertson, E.J., et al., *Design, Synthesis, Assembly, and Engineering of Peptoid Nanosheets.* Accounts of Chemical Research, 2016. **49**(3): p. 379-389.

87. Olivier, G.K., et al., *Antibody-Mimetic Peptoid Nanosheets for Molecular Recognition.* ACS Nano, 2013. **7**(10): p. 9276-9286.

88. Kölmel, D.K., et al., *Cell Penetrating Peptoids (CPPos): Synthesis of a Small Combinatorial Library by Using IRORI MiniKans.* Pharmaceuticals, 2012. **5**(12): p. 1265-1281.

89. Kölmel, D.K., et al., *Cell-penetrating peptoids: Introduction of novel cationic side chains.* European Journal of Medicinal Chemistry, 2014. **79**: p. 231-243.

90. Wender, P.A., et al., *The Design, Synthesis, and Evaluation of Molecules That Enable or Enhance Cellular Uptake: Peptoid Molecular Transporters.* Proceedings of the National Academy of Sciences of the United States of America, 2000. **97**(24): p. 13003-13008.

91. Rönicke, F., *Synthese und in vivo Screening einer Bibliothek zellpenetrierender Peptoide zur Isolation organspezifischer Transportermoleküle.* 2015.

92. Furniss, D., et al., *Peptoids and polyamines going sweet: Modular synthesis of glycosylated peptoids and polyamines using click chemistry.* Beilstein J Org Chem, 2013. **9**: p. 56-63.

93.   Vollrath, S.B., et al., *Amphiphilic peptoid transporters--synthesis and evaluation.* Org Biomol Chem, 2013. **11**(47): p. 8197-201.

94.   Jang, H., et al., *Click to fit: versatile polyvalent display on a peptidomimetic scaffold.* Org Lett, 2005. **7**(10): p. 1951-4.

95.   Birtalan, E., et al., *Investigating rhodamine B-labeled peptoids: scopes and limitations of its applications.* Biopolymers, 2011. **96**(5): p. 694-701.

96.   Huang, W., et al., *Peptoid transporters: effects of cationic, amphipathic structure on their cellular uptake.* Molecular bioSystems, 2012. **8**(10): p. 2626-2628.

97.   Murphy, J.E., et al., *A combinatorial approach to the discovery of efficient cationic peptoid reagents for gene delivery.* Proc Natl Acad Sci U S A, 1998. **95**(4): p. 1517-22.

98.   Kwon, Y.-U. and T. Kodadek, *Quantitative Evaluation of the Relative Cell Permeability of Peptoids and Peptides.* Journal of the American Chemical Society, 2007. **129**(6): p. 1508-1509.

99.   Merrifield, R.B., *Solid Phase Peptide Synthesis. I. The Synthesis of a Tetrapeptide.* Journal of the American Chemical Society, 1963. **85**(14): p. 2149-2154.

100.   Zuckermann, R.N., et al., *Efficient method for the preparation of peptoids [oligo(N-substituted glycines)] by submonomer solid-phase synthesis.* Journal of the American Chemical Society, 1992. **114**(26): p. 10646-10647.

101.   Geysen, H.M., R.H. Meloen, and S.J. Barteling, *Use of peptide synthesis to probe viral antigens for epitopes to a resolution of a single amino acid.* Proceedings of the National Academy of Sciences of the United States of America, 1984. **81**(13): p. 3998-4002.

102.   VALERIO, R.M., et al., *Multipin peptide synthesis at the micromole scale using 2-hydroxyethyl methacryiate grafted polyethylene supports.* International Journal of Peptide and Protein Research, 1993. **42**(1): p. 1-9.

103.   Furka, A., et al., *General method for rapid synthesis of multicomponent peptide mixtures.* Int J Pept Protein Res, 1991. **37**(6): p. 487-93.

104.   Lam, K.S., et al., *A new type of synthetic peptide library for identifying ligand-binding activity.* Nature, 1991. **354**(6348): p. 82-4.

105.   Lam, K.S., M. Lebl, and V. Krchňák, *The "One-Bead-One-Compound" Combinatorial Library Method.* Chemical Reviews, 1997. **97**(2): p. 411-448.

106.   Goodman, B.A., *Managing the Workflow of a High-Throughput Organic Synthesis Laboratory: A Marriage of Automation and Information Management Technologies.* JALA: Journal of the Association for Laboratory Automation, 1999. **4**(6): p. 48-52.

107.   Nicolaou, K.C., et al., *Radiofrequency Encoded Combinatorial Chemistry.* Angewandte Chemie International Edition in English, 1995. **34**(20): p. 2289-2291.

108.   Xiao, X.Y., et al., *Solid-phase combinatorial synthesis using MicroKan reactors, Rf tagging, and directed sorting.* Biotechnology and Bioengineering, 2000. **71**(1): p. 44-50.

109.   Kolb, H.C., M.G. Finn, and K.B. Sharpless, *Click Chemistry: Diverse Chemical Function from a Few Good Reactions.* Angewandte Chemie International Edition, 2001. **40**(11): p. 2004-2021.

110.   Huisgen, R., *1,3-Dipolar Cycloadditions. Past and Future.* Angewandte Chemie International Edition in English, 1963. **2**(10): p. 565-598.

111.   Dondoni, A., *The emergence of thiol-ene coupling as a click process for materials and bioorganic chemistry.* Angew Chem Int Ed Engl, 2008. **47**(47): p. 8995-7.

112.   Wallace, D.C., *Mitochondria and cancer.* Nature Reviews Cancer, 2012. **12**: p. 685.

113.   Mattson, M.P., M. Gleichmann, and A. Cheng, *Mitochondria in Neuroplasticity and Neurological Disorders.* Neuron, 2008. **60**(5): p. 748-766.

114.   Lipinski, C.A., et al., *Experimental and computational approaches to estimate solubility and permeability in drug discovery and development settings1PII of original article: S0169-409X(96)00423-1. The article was originally published in Advanced Drug Delivery Reviews 23 (1997) 3–25.1.* Advanced Drug Delivery Reviews, 2001. **46**(1): p. 3-26.

115.   Ghose, A.K., V.N. Viswanadhan, and J.J. Wendoloski, *A Knowledge-Based Approach in Designing Combinatorial or Medicinal Chemistry Libraries for Drug Discovery. 1. A Qualitative*

and *Quantitative Characterization of Known Drug Databases.* Journal of Combinatorial Chemistry, 1999. **1**(1): p. 55-68.

116. Pajouhesh, H. and G.R. Lenz, *Medicinal chemical properties of successful central nervous system drugs.* NeuroRx : the journal of the American Society for Experimental NeuroTherapeutics, 2005. **2**(4): p. 541-553.

117. *PubChem.* [Database] [cited 2018 15.11.2018]; Open Chemistry Database]. Available from: https://pubchem.ncbi.nlm.nih.gov/.

118. Yousif, L.F., et al., *Mitochondria-Penetrating Peptides: Sequence Effects and Model Cargo Transport.* ChemBioChem, 2009. **10**(12): p. 2081-2088.

119. *Molinspiration.*

120. Herce, H.D., A.E. Garcia, and M.C. Cardoso, *Fundamental Molecular Mechanism for the Cellular Uptake of Guanidinium-Rich Molecules.* Journal of the American Chemical Society, 2014. **136**(50): p. 17459-17467.

121. Danial, N.N. and S.J. Korsmeyer, *Cell death: critical control points.* Cell, 2004. **116**(2): p. 205-19.

122. Newmeyer, D.D. and S. Ferguson-Miller, *Mitochondria: releasing power for life and unleashing the machineries of death.* Cell, 2003. **112**(4): p. 481-90.

123. Gunter, T.E., et al., *Calcium and mitochondria.* FEBS Lett, 2004. **567**(1): p. 96-102.

124. Finkel, T., *Signal transduction by reactive oxygen species.* J Cell Biol, 2011. **194**(1): p. 7-15.

125. Fairlie, D., G. Abenante, and D.R. March, *Macrocyclic peptidomimetics - Forcing peptides into bioactive conformations.* Vol. 2. 1995. 654-686.

126. Baeriswyl, V. and C. Heinis, *Phage selection of cyclic peptide antagonists with increased stability toward intestinal proteases.* Protein Engineering, Design and Selection, 2013. **26**(1): p. 81-89.

127. Joo, S.H., *Cyclic peptides as therapeutic agents and biochemical tools.* Biomolecules and Therapeutics, 2012. **20**(1): p. 19-26.

128. Synge, R.L., *'Gramicidin S': over-all chemical characteristics and amino-acid composition.* Biochem J, 1945. **39**(4): p. 363-7.

129. Horton, D.A., G.T. Bourne, and M.L. Smythe, *Exploring privileged structures: the combinatorial synthesis of cyclic peptides.* J Comput Aided Mol Des, 2002. **16**(5-6): p. 415-30.

130. Perea, S.E., et al., *CIGB-300: A peptide-based drug that impairs the Protein Kinase CK2-mediated phosphorylation.* Semin Oncol, 2018. **45**(1-2): p. 58-67.

131. Culf, A.S., et al., *Small Head-to-Tail Macrocyclic α-Peptoids.* Organic Letters, 2014. **16**(10): p. 2780-2783.

132. Hjelmgaard, T., et al., *Convenient Solution-Phase Synthesis and Conformational Studies of Novel Linear and Cyclic α,β-Alternating Peptoids.* Organic Letters, 2009. **11**(18): p. 4100-4103.

133. Huang, M.L., et al., *Amphiphilic Cyclic Peptoids That Exhibit Antimicrobial Activity by Disrupting Staphylococcus aureus Membranes.* European Journal of Organic Chemistry, 2013. **2013**(17): p. 3560-3566.

134. Khan, S.N., et al., *Ring-Closing Metathesis Approaches for the Solid-Phase Synthesis of Cyclic Peptoids.* Organic Letters, 2011. **13**(7): p. 1582-1585.

135. Shin, S.B.Y., et al., *Cyclic Peptoids.* Journal of the American Chemical Society, 2007. **129**(11): p. 3218-3225.

136. Kwon, Y.U. and T. Kodadek, *Encoded combinatorial libraries for the construction of cyclic peptoid microarrays.* Chem Commun (Camb), 2008(44): p. 5704-6.

137. Holub, J.M., H. Jang, and K. Kirshenbaum, *Fit To Be Tied: Conformation-Directed Macrocyclization of Peptoid Foldamers.* Organic Letters, 2007. **9**(17): p. 3275-3278.

138. Chen, X., et al., *Platinum-based agents for individualized cancer treatment.* Curr Mol Med, 2013. **13**(10): p. 1603-12.

139. Galmarini, C.M., J.R. Mackey, and C. Dumontet, *Nucleoside analogues and nucleobases in cancer treatment.* Lancet Oncol, 2002. **3**(7): p. 415-24.

140. Lamanna, N. and M. Weiss, *Purine analogs in leukemia.* Adv Pharmacol, 2004. **51**: p. 107-25.

141. Lee, Y.-C., et al., *Targeting of Topoisomerase I for Prognoses and Therapeutics of Camptothecin-Resistant Ovarian Cancer.* PloS one, 2015. **10**(7): p. e0132579-e0132579.

142. Ohira, M., et al., *A novel anti-microtubule agent with carbazole and benzohydrazide structures suppresses tumor cell growth in vivo.* Biochim Biophys Acta, 2015. **1850**(9): p. 1676-84.

143. Iwamoto, T., *Clinical application of drug delivery systems in cancer chemotherapy: review of the efficacy and side effects of approved drugs.* Biol Pharm Bull, 2013. **36**(5): p. 715-8.

144. Kruh, G.D. and L.J. Goldstein, *Doxorubicin and multidrug resistance.* Curr Opin Oncol, 1993. **5**(6): p. 1029-34.

145. Liscovitch, M. and Y. Lavie, *Cancer multidrug resistance: a review of recent drug discovery research.* IDrugs, 2002. **5**(4): p. 349-55.

146. Faller, B.A., V.G. Robu, and H. Borghaei, *Therapy-related acute myelogenous leukemia with an 11q23/MLL translocation following adjuvant cisplatin and vinorelbine for non-small-cell lung cancer.* Clin Lung Cancer, 2009. **10**(6): p. 438-40.

147. van Leeuwen, F.E., *Risk of acute myelogenous leukaemia and myelodysplasia following cancer treatment.* Baillieres Clin Haematol, 1996. **9**(1): p. 57-85.

148. Aidan, J.C., N.R. Priddee, and J.J. McAleer, *Chemotherapy causes cancer! A case report of therapy related acute myeloid leukaemia in early stage breast cancer.* The Ulster medical journal, 2013. **82**(2): p. 97-99.

149. Gaspar, D., A.S. Veiga, and M.A.R.B. Castanho, *From antimicrobial to anticancer peptides. A review.* Frontiers in microbiology, 2013. **4**: p. 294-294.

150. Iwasaki, T., et al., *Selective cancer cell cytotoxicity of enantiomeric 9-mer peptides derived from beetle defensins depends on negatively charged phosphatidylserine on the cell surface.* Peptides, 2009. **30**(4): p. 660-8.

151. Papo, N. and Y. Shai, *Host defense peptides as new weapons in cancer treatment.* Cell Mol Life Sci, 2005. **62**(7-8): p. 784-90.

152. Storz, P., *Reactive oxygen species in tumor progression.* Front Biosci, 2005. **10**: p. 1881-96.

153. Liou, G.Y. and P. Storz, *Reactive oxygen species in cancer.* Free Radic Res, 2010. **44**(5): p. 479-96.

154. Szatrowski, T.P. and C.F. Nathan, *Production of large amounts of hydrogen peroxide by human tumor cells.* Cancer Res, 1991. **51**(3): p. 794-8.

155. Auten, R.L. and J.M. Davis, *Oxygen toxicity and reactive oxygen species: the devil is in the details.* Pediatr Res, 2009. **66**(2): p. 121-7.

156. Zou, Z.Z., et al., *Synergistic induction of apoptosis by salinomycin and gefitinib through lysosomal and mitochondrial dependent pathway overcomes gefitinib resistance in colorectal cancer.* Oncotarget, 2017. **8**(14): p. 22414-22432.

157. Eruslanov, E. and S. Kusmartsev, *Identification of ROS using oxidized DCFDA and flow-cytometry.* Methods Mol Biol, 2010. **594**: p. 57-72.

158. Cukierman, E., R. Pankov, and K.M. Yamada, *Cell interactions with three-dimensional matrices.* Current Opinion in Cell Biology, 2002. **14**(5): p. 633-639.

159. Yamada, K.M. and E. Cukierman, *Modeling Tissue Morphogenesis and Cancer in 3D.* Cell, 2007. **130**(4): p. 601-610.

160. Hackam, D.G. and D.A. Redelmeier, *Translation of research evidence from animals to humans.* JAMA, 2006. **296**(14): p. 1727-1732.

161. Costa, E.C., et al., *3D tumor spheroids: an overview on the tools and techniques used for their analysis.* Biotechnology Advances, 2016. **34**(8): p. 1427-1441.

162. Trédan, O., et al., *Drug Resistance and the Solid Tumor Microenvironment.* JNCI: Journal of the National Cancer Institute, 2007. **99**(19): p. 1441-1454.

163. Breslin, S. and L. O'Driscoll, *Three-dimensional cell culture: the missing link in drug discovery.* Drug Discovery Today, 2013. **18**(5): p. 240-249.

164. Kriston-Vizi, J. and H. Flotow, *Getting the whole picture: High content screening using three-dimensional cellular model systems and whole animal assays.* Cytometry Part A, 2017. **91**(2): p. 152-159.

165. Correa de Sampaio, P., et al., *A Heterogeneous In Vitro Three Dimensional Model of Tumour-Stroma Interactions Regulating Sprouting Angiogenesis.* PLOS ONE, 2012. **7**(2): p. e30753.

166. Costa, E.C., et al., *Optimization of liquid overlay technique to formulate heterogenic 3D co-cultures models.* Biotechnology and Bioengineering, 2014. **111**(8): p. 1672-1685.

167. Kappings, V., et al., *vasQchip: A Novel Microfluidic, Artificial Blood Vessel Scaffold for Vascularized 3D Tissues.* Advanced Materials Technologies, 2018. **3**(4): p. 1700246.

168. Kappings, V., *Tumormodelle in vitro: von 2D-Zellkulturen zum durchbluteten Organ-on-a-chip System.* 2017.

169. Bradford, Y.M., et al., *Zebrafish Models of Human Disease: Gaining Insight into Human Disease at ZFIN.* ILAR journal, 2017. **58**(1): p. 4-16.

170. Howe, K., et al., *The zebrafish reference genome sequence and its relationship to the human genome.* Nature, 2013. **496**(7446): p. 498-503.

171. Lammer, E., et al., *Is the fish embryo toxicity test (FET) with the zebrafish (Danio rerio) a potential alternative for the fish acute toxicity test?* Comparative Biochemistry and Physiology Part C: Toxicology & Pharmacology, 2009. **149**(2): p. 196-209.

172. Henn, K. and T. Braunbeck, *Dechorionation as a tool to improve the fish embryo toxicity test (FET) with the zebrafish (Danio rerio).* Comp Biochem Physiol C Toxicol Pharmacol, 2011. **153**(1): p. 91-8.

173. White, R.M., et al., *Transparent adult zebrafish as a tool for in vivo transplantation analysis.* Cell stem cell, 2008. **2**(2): p. 183-189.

174. Peravali, R., et al., *Automated feature detection and imaging for high-resolution screening of zebrafish embryos.* BioTechniques, 2011. **50**(5): p. 319-324.

175. Sidi, S., et al., *Chk1 suppresses a caspase-2 apoptotic response to DNA damage that bypasses p53, Bcl-2, and caspase-3.* Cell, 2008. **133**(5): p. 864-77.

176. Greiling, T.M. and J.I. Clark, *Early lens development in the zebrafish: a three-dimensional time-lapse analysis.* Dev Dyn, 2009. **238**(9): p. 2254-65.

177. Bibliowicz, J., R.K. Tittle, and J.M. Gross, *Chapter 7 - Toward a Better Understanding of Human Eye Disease: Insights From the Zebrafish, Danio rerio*, in *Progress in Molecular Biology and Translational Science*, K.T. Chang and K.-T. Min, Editors. 2011, Academic Press. p. 287-330.

178. Fadool, J.M. and J.E. Dowling, *Zebrafish: A model system for the study of eye genetics.* Progress in Retinal and Eye Research, 2008. **27**(1): p. 89-110.

179. Schmitt, E.A. and J.E. Dowling, *Early retinal development in the zebrafish, Danio rerio: Light and electron microscopic analyses.* Journal of Comparative Neurology, 1999. **404**(4): p. 515-536.

180. Wallace, K.N. and M. Pack, *Unique and conserved aspects of gut development in zebrafish.* Developmental Biology, 2003. **255**(1): p. 12-29.

181. Wallace, K.N., et al., *Intestinal growth and differentiation in zebrafish.* Mechanisms of Development, 2005. **122**(2): p. 157-173.

182. Lawson, N.D. and B.M. Weinstein, *Arteries and veins: making a difference with zebrafish.* Nature Reviews Genetics, 2002. **3**: p. 674.

183. Drummond, I.A. and A.J. Davidson, *Zebrafish kidney development.* Methods Cell Biol, 2016. **134**: p. 391-429.

184. Swanhart, L.M., et al., *Zebrafish kidney development: basic science to translational research.* Birth defects research. Part C, Embryo today : reviews, 2011. **93**(2): p. 141-156.

185. Chou, C.-W., et al., *The endoderm indirectly influences morphogenetic movements of the zebrafish head kidney through the posterior cardinal vein and VegfC.* Scientific Reports, 2016. **6**: p. 30677.

186. Raible, D.W. and G.J. Kruse, *Organization of the lateral line system in embryonic zebrafish.* J Comp Neurol, 2000. **421**(2): p. 189-98.

187.   Hara, T.J., *The diversity of chemical stimulation in fish olfaction and gustation.* Reviews in Fish Biology and Fisheries, 1994. **4**(1): p. 1-35.
188.   Laberge, F. and T.J. Hara, *Neurobiology of fish olfaction: a review.* Brain Research Reviews, 2001. **36**(1): p. 46-59.
189.   Saxena, A., B.N. Peng, and M.E. Bronner, *Sox10-dependent neural crest origin of olfactory microvillous neurons in zebrafish.* eLife, 2013. **2**: p. e00336.
190.   Mistry, A., S. Stolnik, and L. Illum, *Nanoparticles for direct nose-to-brain delivery of drugs.* International Journal of Pharmaceutics, 2009. **379**(1): p. 146-157.
191.   Illum, L., *Transport of drugs from the nasal cavity to the central nervous system.* European Journal of Pharmaceutical Sciences, 2000. **11**(1): p. 1-18.
192.   Thorne, R.G., et al., *Quantitative analysis of the olfactory pathway for drug delivery to the brain.* Brain Res, 1995. **692**(1-2): p. 278-82.
193.   Sabin, A.B. and P.K. Olitsky, *The olfactory bulbs in experimental poliomyelitis: Their pathologic condition as an indicator of the portal of entry of the virus.* Journal of the American Medical Association, 1937. **108**(1): p. 21-24.
194.   Sabin, A.B. and P.K. Olitsky, *INFLUENCE OF HOST FACTORS ON NEUROINVASIVENESS OF VESICULAR STOMATITIS VIRUS.* I. EFFECT OF AGE ON THE INVASION OF THE BRAIN BY VIRUS INSTILLED IN THE NOSE, 1937. **66**(1): p. 15-34.
195.   Landsteiner, K. and C. Levaditi, *Étude expérimentale de la poliomyélite aiguë (maladie de Heine-Medin).* 1910, Paris.
196.   Flexner, S., *The mode of infection in epidemic poliomyelitis.* Journal of the American Medical Association, 1912. **LIX**(15): p. 1371-1372.
197.   Hastings, L. and J.E. Evans, *Olfactory primary neurons as a route of entry for toxic agents into the CNS.* NeuroToxicology, 1991. **12**(4): p. 707-714.
198.   Gopinath, P.G., G. Gopinath, and T.C.A. Kumar, *Target site of intranasally sprayed substances and their transport across the nasal mucosa: A new insight into the intranasal route of drug-delivery.* Current Therapeutic Research - Clinical and Experimental, 1978. **23**(5 I): p. 596-607.
199.   Kristensson, K. and Y. Olsson, *Uptake of exogenous proteins in mouse olfactory cells.* Acta Neuropathologica, 1971. **19**(2): p. 145-154.
200.   Whitfield, T.T., *Shedding new light on the origins of olfactory neurons.* eLife, 2013. **2**: p. e00648.
201.   Graff, C.L. and G.M. Pollack, *Nasal drug administration: potential for targeted central nervous system delivery.* Journal of Pharmaceutical Sciences, 2005. **94**(6): p. 1187-1195.
202.   Mathison, S., R. Nagilla, and U.B. Kompella, *Nasal route for direct delivery of solutes to the central nervous system: fact or fiction?* J Drug Target, 1998. **5**(6): p. 415-41.
203.   Popova, A.A., et al., *Fish-Microarray: A Miniaturized Platform for Single-Embryo High-Throughput Screenings.* Advanced Functional Materials, 2018. **28**(3): p. 1703486.
204.   Popova, A.A., et al., *Droplet-Array (DA) Sandwich Chip: A Versatile Platform for High-Throughput Cell Screening Based on Superhydrophobic-Superhydrophilic Micropatterning.* Adv Mater, 2015. **27**(35): p. 5217-22.
205.   Yang, Z., et al., *Recent advances in organic thermally activated delayed fluorescence materials.* Chem Soc Rev, 2017. **46**(3): p. 915-1016.
206.   Salazar, F.A., A. Fedorov, and M.N. Berberan-Santos, *A study of thermally activated delayed fluorescence in C60.* Chemical Physics Letters, 1997. **271**(4): p. 361-366.
207.   Münch, S.W., *Festphasensynthese neuartiger molekularer Transporter und funktioneller Peptide.* 2017.
208.   Feuser, P.E., et al., *Encapsulation of magnetic nanoparticles in poly(methyl methacrylate) by miniemulsion and evaluation of hyperthermia in U87MG cells.* European Polymer Journal, 2015. **68**: p. 355-365.

209. Behzadi, S., et al., *Cellular uptake of nanoparticles: journey inside the cell.* Chem Soc Rev, 2017. **46**(14): p. 4218-4244.

210. Singh, R. and J.W. Lillard, Jr., *Nanoparticle-based targeted drug delivery.* Experimental and molecular pathology, 2009. **86**(3): p. 215-223.

211. Wilczewska, A.Z., et al., *Nanoparticles as drug delivery systems.* Pharmacological Reports, 2012. **64**(5): p. 1020-1037.

212. Savjani, K.T., A.K. Gajjar, and J.K. Savjani, *Drug Solubility: Importance and Enhancement Techniques.* ISRN Pharmaceutics, 2012. **2012**: p. 195727.

213. Hawkins, M.J., P. Soon-Shiong, and N. Desai, *Protein nanoparticles as drug carriers in clinical medicine.* Adv Drug Deliv Rev, 2008. **60**(8): p. 876-85.

214. Loadman, P.M., et al., *Pharmacokinetics of PK1 and Doxorubicin in Experimental Colon Tumor Models with Differing Responses to PK1.* Clinical Cancer Research, 1999. **5**(11): p. 3682-3688.

215. Vauthier, C., et al., *Drug delivery to resistant tumors: the potential of poly(alkyl cyanoacrylate) nanoparticles.* J Control Release, 2003. **93**(2): p. 151-60.

216. Pepin, X., et al., *On the use of ion-pair chromatography to elucidate doxorubicin release mechanism from polyalkylcyanoacrylate nanoparticles at the cellular level.* J Chromatogr B Biomed Sci Appl, 1997. **702**(1-2): p. 181-91.

217. Heiler, C., et al., *Photochemically Induced Folding of Single Chain Polymer Nanoparticles in Water.* ACS Macro Letters, 2017. **6**(1): p. 56-61.

218. Heiler, C., *Photo-induced formation of fluorescent single chain polymer nanoparticles as imaging agents in biology.* 2018.

219. Alshatwi, A.A., et al., *Al(2)O(3) nanoparticles induce mitochondria-mediated cell death and upregulate the expression of signaling genes in human mesenchymal stem cells.* J Biochem Mol Toxicol, 2012. **26**(11): p. 469-76.

220. Guan, R., et al., *Cytotoxicity, oxidative stress, and genotoxicity in human hepatocyte and embryonic kidney cells exposed to ZnO nanoparticles.* Nanoscale Res Lett, 2012. **7**(1): p. 602.

221. Patra, H.K., et al., *Cell selective response to gold nanoparticles.* Nanomedicine: Nanotechnology, Biology and Medicine, 2007. **3**(2): p. 111-119.

222. Horton, K.L., et al., *Mitochondria-Penetrating Peptides.* Chemistry & Biology, 2008. **15**(4): p. 375-382.

223. Raldúa, D., et al., *Zebrafish as a Vertebrate Model to Assess Sublethal Effects and Health Risks of Emerging Pollutants.* 2012.

224. Alexandre, D. and A. Ghysen, *Somatotopy of the lateral line projection in larval zebrafish.* Proceedings of the National Academy of Sciences of the United States of America, 1999. **96**(13): p. 7558-7562.

225. Collazo, A., S.E. Fraser, and P.M. Mabee, *A dual embryonic origin for vertebrate mechanoreceptors.* Science, 1994. **264**(5157): p. 426-30.

226. Hosoya, K., et al., *Lipophilicity and transporter influence on blood-retinal barrier permeability: a comparison with blood-brain barrier permeability.* Pharm Res, 2010. **27**(12): p. 2715-24.

227. Kwon, Y.-U. and T. Kodadek, *Quantitative Comparison of the Relative Cell Permeability of Cyclic and Linear Peptides.* Chemistry & Biology, 2007. **14**(6): p. 671-677.

228. Nakatsu, M.N., J. Davis, and C.C.W. Hughes, *Optimized fibrin gel bead assay for the study of angiogenesis.* Journal of visualized experiments : JoVE, 2007(3): p. 186-186.

229. Mikut, R., *Computer-Based Analysis, Visualization, and Interpretation of Antimicrobial Peptide Activities*, in *Antimicrobial Peptides: Methods and Protocols*, A. Giuliani and A.C. Rinaldi, Editors. 2010, Humana Press: Totowa, NJ. p. 287-299.

230. Abraham, M.J., et al., *GROMACS: High performance molecular simulations through multi-level parallelism from laptops to supercomputers.* SoftwareX, 2015. **1-2**: p. 19-25.

231. Berendsen, H.J.C., D. van der Spoel, and R. van Drunen, *GROMACS: A message-passing parallel molecular dynamics implementation.* Computer Physics Communications, 1995. **91**(1): p. 43-56.

232.   Wang, J., et al., *Automatic atom type and bond type perception in molecular mechanical calculations.* J Mol Graph Model, 2006. **25**(2): p. 247-60.
233.   Wang, J., et al., *Development and testing of a general amber force field.* J Comput Chem, 2004. **25**(9): p. 1157-74.
234.   *AmberTools 14.* 2014.
235.   Frisch, M.J., et al., *Gaussian 16 Rev. B.01.* 2016: Wallingford, CT.
236.   Singh, U.C. and P.A. Kollman, *An approach to computing electrostatic charges for molecules.* Journal of Computational Chemistry, 1984. **5**(2): p. 129-145.
237.   Besler, B.H., K.M. Merz, and P.A. Kollman, *Atomic charges derived from semiempirical methods.* Journal of Computational Chemistry, 1990. **11**(4): p. 431-439.
238.   Petersson, G.A., et al., *A complete basis set model chemistry. I. The total energies of closed-shell atoms and hydrides of the first-row elements.* The Journal of Chemical Physics, 1988. **89**(4): p. 2193-2218.
239.   Petersson, G.A. and M.A. Al-Laham, *A complete basis set model chemistry. II. Open-shell systems and the total energies of the first-row atoms.* The Journal of Chemical Physics, 1991. **94**(9): p. 6081-6090.
240.   Evans, D.J. and B.L. Holian, *The Nose–Hoover thermostat.* The Journal of Chemical Physics, 1985. **83**(8): p. 4069-4074.
241.   Parrinello, M. and A. Rahman, *Polymorphic transitions in single crystals: A new molecular dynamics method.* Journal of Applied Physics, 1981. **52**(12): p. 7182-7190.

8. Appendix

Table 18: Peptoids in library 1. Peptoid numbers, sequences and respective molecular weights (g/mol) are shown. Success of synthesis was verified by MALDI TOF mass spectrometry and identified peptoids are marked with "x".

#	Sequence	MW		#	Sequence	MW	
P1	Nprg-Nprg-Nprg-Nprg	822.49	x	P44	Nprg-Mys-Mys-Nphe	940.62	x
P2	Nprg-Nprg-Nprg-Npcb	908.48	0	P45	Nprg-Mys-Nphe-Nprg	907.57	x
P3	Nprg-Nprg-Nprg-Mys	855.55	x	P46	Nprg-Mys-Nphe-Npcb	993.56	x
P4	Nprg-Nprg-Nprg-Nphe	874.52	x	P47	Nprg-Mys-Nphe-Mys	940.62	x
P5	Nprg-Nprg-Npcb-Nprg	908.48	x	P48	Nprg-Mys-Nphe-Nphe	959.60	x
P6	Nprg-Nprg-Npcb-Npcb	994.47	x	P49	Nprg-Nphe-Nprg-Nprg	874.52	x
P7	Nprg-Nprg-Npcb-Mys	941.53	0	P50	Nprg-Nphe-Nprg-Npcb	960.51	x
P8	Nprg-Nprg-Npcb-Nphe	960.51	x	P51	Nprg-Nphe-Nprg-Mys	907.57	x
P9	Nprg-Nprg-Mys-Nprg	855.54	x	P52	Nprg-Nphe-Nprg-Nphe	926.55	x
P10	Nprg-Nprg-Mys-Npcb	941.53	x	P53	Nprg-Nphe-Npcb-Nprg	960.51	0
P11	Nprg-Nprg-Mys-Mys	888.59	x	P54	Nprg-Nphe-Npcb-Npcb	1046.5	x
P12	Nprg-Nprg-Mys-Nphe	907.57	x	P55	Nprg-Nphe-Npcb-Mys	993.56	x
P13	Nprg-Nprg-Nphe-Nprg	874.52	x	P56	Nprg-Nphe-Npcb-Nphe	1012.54	x
P14	Nprg-Nprg-Nphe-Npcb	960.51	x	P57	Nprg-Nphe-Mys-Nprg	907.57	0
P15	Nprg-Nprg-Nphe-Mys	907.57	x	P58	Nprg-Nphe-Mys-Npcb	993.56	x
P16	Nprg-Nprg-Nphe-Nphe	926.55	0	P59	Nprg-Nphe-Mys-Mys	940.62	x
P17	Nprg-Npcb-Nprg-Nprg	908.48	x	P60	Nprg-Nphe-Mys-Nphe	959.60	x
P18	Nprg-Npcb-Nprg-Npcb	994.47	x	P61	Nprg-Nphe-Nphe-Nprg	926.55	x
P19	Nprg-Npcb-Nprg-Mys	941.53	x	P62	Nprg-Nphe-Nphe-Npcb	1012.54	x
P20	Nprg-Npcb-Nprg-Nphe	960.51	x	P63	Nprg-Nphe-Nphe-Mys	959.60	x
P21	Nprg-Npcb-Npcb-Nprg	994.47	x	P64	Nprg-Nphe-Nphe-Nphe	978.58	x
P22	Nprg-Npcb-Npcb-Npcb	1080.46	x	P65	Npcb-Nprg-Nprg-Nprg	908.48	x
P23	Nprg-Npcb-Npcb-Mys	1029.52	x	P66	Npcb-Nprg-Nprg-Npcb	994.47	x
P24	Nprg-Npcb-Npcb-Nphe	1046.50	x	P67	Npcb-Nprg-Nprg-Mys	941.53	x
P25	Nprg-Npcb-Mys-Nprg	941.53	x	P68	Npcb-Nprg-Nprg-Nphe	960.51	x
P26	Nprg-Npcb-Mys-Npcb	1027.52	x	P69	Npcb-Nprg-Npcb-Nprg	994.51	x
P27	Nprg-Npcb-Mys-Mys	974.58	x	P70	Npcb-Nprg-Npcb-Npcb	1080.46	x
P28	Nprg-Npcb-Mys-Nphe	993.56	x	P71	Npcb-Nprg-Npcb-Mys	1027.52	x
P29	Nprg-Npcb-Nphe-Nprg	960.51	x	P72	Npcb-Nprg-Npcb-Nphe	1046.50	x
P30	Nprg-Npcb-Nphe-Npcb	1046.50	x	P73	Npcb-Nprg-Mys-Nprg	941.53	x
P31	Nprg-Npcb-Nphe-Mys	993.56	x	P74	Npcb-Nprg-Mys-Npcb	1027.52	x
P32	Nprg-Npcb-Nphe-Nphe	1012.54	0	P75	Npcb-Nprg-Mys-Mys	974.58	x
P33	Nprg-Mys-Nprg-Nprg	855.54	x	P76	Npcb-Nprg-Mys-Nphe	993.56	x
P34	Nprg-Mys-Nprg-Npcb	941.53	0	P77	Npcb-Nprg-Nphe-Nprg	960.51	x
P35	Nprg-Mys-Nprg-Mys	888.59	x	P78	Npcb-Nprg-Nphe-Npcb	1046.5	0
P36	Nprg-Mys-Nprg-Nphe	907.57	x	P79	Npcb-Nprg-Nphe-Mys	993.56	x
P37	Nprg-Mys-Npcb-Nprg	941.53	x	P80	Npcb-Nprg-Nphe-Nphe	1012.54	x
P38	Nprg-Mys-Npcb-Npcb	1027.52	x	P81	Npcb-Npcb-Nprg-Nprg	994.47	x
P39	Nprg-Mys-Npcb-Mys	974.58	x	P82	Npcb-Npcb-Nprg-Npcb	1080.46	x
P40	Nprg-Mys-Npcb-Nphe	993.56	X	P83	Npcb-Npcb-Nprg-Mys	1027.52	x
P41	Nprg-Mys-Mys-Nprg	888.59	x	P84	Npcb-Npcb-Nprg-Nphe	1046.50	x
P42	Nprg-Mys-Mys-Npcb	974.58	x	P85	Npcb-Npcb-Npcb-Nprg	1080.46	x
P43	Nprg-Mys-Mys-Mys	921.64	0	P86	Npcb-Npcb-Npcb-Npcb	1166.45	x
P87	Npcb-Npcb-Npcb-Mys	1113.51	x	P134	Mys-Nprg-Npcb-Npcb	1027.52	0
P88	Npcb-Npcb-Npcb-Nphe	1132.49	x	P135	Mys-Nprg-Npcb-Mys	974.58	x

P89	Npcb-Npcb-Mys-Nprg	1027.52	x	P136	Mys-Nprg-Npcb-Nphe	993.56	x
P90	Npcb-Npcb-Mys-Npcb	1113.51	x	P137	Mys-Nprg-Mys-Nprg	888.59	x
P91	Npcb-Npcb-Mys-Mys	1060.57	x	P138	Mys-Nprg-Mys-Npcb	974.58	x
P92	Npcb-Npcb-Mys-Nphe	1079.55	x	P139	Mys-Nprg-Mys-Mys	921.64	x
P93	Npcb-Npcb-Nphe-Nprg	1046.50	x	P140	Mys-Nprg-Mys-Nphe	940.62	x
P94	Npcb-Npcb-Nphe-Npcb	1132.49	x	P141	Mys-Nprg-Nphe-Nprg	907.57	x
P95	Npcb-Npcb-Nphe-Mys	1079.55	x	P142	Mys-Nprg-Nphe-Npcb	993.56	x
P96	Npcb-Npcb-Nphe-Nphe	1098.53	x	P143	Mys-Nprg-Nphe-Mys	940.62	0
P97	Npcb-Mys-Nprg-Nprg	941.53	0	P144	Mys-Nprg-Nphe-Nphe	959.6	x
P98	Npcb-Mys-Nprg-Npcb	1027.52	x	P145	Mys-Npcb-Nprg-Nprg	941.53	x
P99	Npcb-Mys-Nprg-Mys	974.58	x	P146	Mys-Npcb-Nprg-Npcb	1027.52	x
P100	Npcb-Mys-Nprg-Nphe	993.56	x	P147	Mys-Npcb-Nprg-Mys	974.58	x
P101	Npcb-Mys-Npcb-Nprg	1027.52	x	P148	Mys-Npcb-Nprg-Nphe	993.56	x
P102	Npcb-Mys-Npcb-Npcb	1113.51	x	P149	Mys-Npcb-Npcb-Nprg	1027.52	x
P103	Npcb-Mys-Npcb-Mys	1060.57	x	P150	Mys-Npcb-Npcb-Npcb	1113.51	x
P104	Npcb-Mys-Npcb-Nphe	1079.55	x	P151	Mys-Npcb-Npcb-Mys	1060.57	x
P105	Npcb-Mys-Mys-Nprg	974.58	x	P152	Mys-Npcb-Npcb-Nphe	1079.55	x
P106	Npcb-Mys-Mys-Npcb	1060.57	x	P153	Mys-Npcb-Mys-Nprg	974.58	x
P107	Npcb-Mys-Mys-Mys	1007.63	x	P154	Mys-Npcb-Mys-Npcb	1060.57	x
P108	Npcb-Mys-Mys-Nphe	1026.61	0	P155	Mys-Npcb-Mys-Mys	1007.63	x
P109	Npcb-Mys-Nphe-Nprg	993.56	x	P156	Mys-Npcb-Mys-Nphe	1026.61	x
P110	Npcb-Mys-Nphe-Npcb	1079.55	x	P157	Mys-Npcb-Nphe-Nprg	993.56	x
P111	Npcb-Mys-Nphe-Mys	1026.61	x	P158	Mys-Npcb-Nphe-Npcb	1079.55	x
P112	Npcb-Mys-Nphe-Nphe	1045.59	x	P159	Mys-Npcb-Nphe-Mys	1026.61	x
P113	Npcb-Nphe-Nprg-Nprg	960.51	x	P160	Mys-Npcb-Nphe-Nphe	1045.59	x
P114	Npcb-Nphe-Nprg-Npcb	1046.5	x	P161	Mys-Mys-Nprg-Nprg	888.59	x
P115	Npcb-Nphe-Nprg-Mys	993.56	x	P162	Mys-Mys-Nprg-Npcb	9974.58	0
P116	Npcb-Nphe-Nprg-Nphe	1012.54	x	P163	Mys-Mys-Nprg-Mys	921.64	x
P117	Npcb-Nphe-Npcb-Nprg	1046.5	x	P164	Mys-Mys-Nprg-Nphe	940.62	0
P118	Npcb-Nphe-Npcb-Npcb	1132.49	x	P165	Mys-Mys-Npcb-Nprg	974.58	x
P119	Npcb-Nphe-Npcb-Mys	1079.55	x	P166	Mys-Mys-Npcb-Npcb	1060.57	x
P120	Npcb-Nphe-Npcb-Nphe	1098.53	x	P167	Mys-Mys-Npcb-Mys	1007.63	x
P121	Npcb-Nphe-Mys-Nprg	993.56	x	P168	Mys-Mys-Npcb-Nphe	1026.61	x
P122	Npcb-Nphe-Mys-Npcb	1079.55	x	P169	Mys-Mys-Mys-Nprg	921.64	x
P123	Npcb-Nphe-Mys-Mys	1026.61	x	P170	Mys-Mys-Mys-Npcb	1007.63	x
P124	Npcb-Nphe-Mys-Nphe	1045.59	x	P171	Mys-Mys-Mys-Mys	954.69	x
P125	Npcb-Nphe-Nphe-Nprg	1012.54	x	P172	Mys-Mys-Mys-Nphe	973.667	x
P126	Npcb-Nphe-Nphe-Npcb	1098.53	x	P173	Mys-Mys-Nphe-Nprg	940.62	0
P127	Npcb-Nphe-Nphe-Mys	1045.59	x	P174	Mys-Mys-Nphe-Npcb	1026.61	x
P128	Npcb-Nphe-Nphe-Nphe	1064.57	x	P175	Mys-Mys-Nphe-Mys	973.67	x
P129	Mys-Nprg-Nprg-Nprg	855.54	x	P176	Mys-Mys-Nphe-Nphe	992.65	x
P130	Mys-Nprg-Nprg-Npcb	941.53	x	P177	Mys-Nphe-Nprg-Nprg	907.57	x
P131	Mys-Nprg-Nprg-Mys	888.59	x	P178	Mys-Nphe-Nprg-Npcb	993.56	x
P132	Mys-Nprg-Nprg-Nphe	907.57	x	P179	Mys-Nphe-Nprg-Mys	940.62	x
P133	Mys-Nprg-Npcb-Nprg	941.53	x	P180	Mys-Nphe-Nprg-Nphe	959.60	x
P181	Mys-Nphe-Npcb-Nprg	993.56	x	P221	Nphe-Npcb-Nphe-Nprg	1012.54	x
P182	Mys-Nphe-Npcb-Npcb	7079.55	x	P222	Nphe-Npcb-Nphe-Npcb	1098.53	0
P183	Mys-Nphe-Npcb-Mys	1026.61	x	P223	Nphe-Npcb-Nphe-Mys	1045.59	x
P184	Mys-Nphe-Npcb-Nphe	1045.59	x	P224	Nphe-Npcb-Nphe-Nphe	1064.57	x
P185	Mys-Nphe-Mys-Nprg	940.62	0	P225	Nphe-Mys-Nprg-Nprg	907.57	x

160

P186	Mys-Nphe-Mys-Npcb	1026.61	x
P187	Mys-Nphe-Mys-Mys	973.67	x
P188	Mys-Nphe-Mys-Nphe	992.65	x
P189	Mys-Nphe-Nphe-Nprg	959.60	x
P190	Mys-Nphe-Nphe-Npcb	1045.59	x
P191	Mys-Nphe-Nphe-Mys	992.65	x
P192	Mys-Nphe-Nphe-Nphe	1011.63	x
P193	Nphe-Nprg-Nprg-Nprg	874.52	x
P194	Nphe-Nprg-Nprg-Npcb	960.51	x
P195	Nphe-Nprg-Nprg-Mys	907.57	x
P196	Nphe-Nprg-Nprg-Nphe	926.55	0
P197	Nphe-Nprg-Npcb-Nprg	960.51	x
P198	Nphe-Nprg-Npcb-Npcb	1046.5	x
P199	Nphe-Nprg-Npcb-Mys	993.56	x
P200	Nphe-Nprg-Npcb-Nphe	1012.54	x
P201	Nphe-Nprg-Mys-Nprg	907.57	x
P202	Nphe-Nprg-Mys-Npcb	993.56	x
P203	Nphe-Nprg-Mys-Mys	940.62	x
P204	Nphe-Nprg-Mys-Nphe	959.60	x
P205	Nphe-Nprg-Nphe-Nprg	926.55	x
P206	Nphe-Nprg-Nphe-Npcb	1012.54	x
P207	Nphe-Nprg-Nphe-Mys	959.60	x
P208	Nphe-Nprg-Nphe-Nphe	978.58	x
P209	Nphe-Npcb-Nprg-Nprg	960.51	0
P210	Nphe-Npcb-Nprg-Npcb	1046.50	x
P211	Nphe-Npcb-Nprg-Mys	993.56	x
P212	Nphe-Npcb-Nprg-Nphe	1012.54	x
P213	Nphe-Npcb-Npcb-Nprg	1046.50	0
P214	Nphe-Npcb-Npcb-Npcb	1132.49	x
P215	Nphe-Npcb-Npcb-Mys	1079.55	x
P216	Nphe-Npcb-Npcb-Nphe	1098.53	x
P217	Nphe-Npcb-Mys-Nprg	993.56	x
P218	Nphe-Npcb-Mys-Npcb	1079.55	x
P219	Nphe-Npcb-Mys-Mys	1026.61	x
P220	Nphe-Npcb-Mys-Nphe	1045.59	x

P226	Nphe-Mys-Nprg-Npcb	993.56	x
P227	Nphe-Mys-Nprg-Mys	940.62	x
P228	Nphe-Mys-Nprg-Nphe	959.60	x
P229	Nphe-Mys-Npcb-Nprg	993.56	x
P230	Nphe-Mys-Npcb-Npcb	1079.55	x
P231	Nphe-Mys-Npcb-Mys	1026.61	x
P232	Nphe-Mys-Npcb-Nphe	1045.59	x
P233	Nphe-Mys-Mys-Nprg	940.62	x
P234	Nphe-Mys-Mys-Npcb	1026.61	x
P235	Nphe-Mys-Mys-Mys	973.67	x
P236	Nphe-Mys-Mys-Nphe	992.65	x
P237	Nphe-Mys-Nphe-Nprg	959.60	x
P238	Nphe-Mys-Nphe-Npcb	1045.59	x
P239	Nphe-Mys-Nphe-Mys	992.65	x
P240	Nphe-Mys-Nphe-Nphe	1011.63	x
P241	Nphe-Nphe-Nprg-Nprg	926.55	x
P242	Nphe-Nphe-Nprg-Npcb	1012.54	x
P243	Nphe-Nphe-Nprg-Mys	959.60	x
P244	Nphe-Nphe-Nprg-Nphe	978.58	x
P245	Nphe-Nphe-Npcb-Nprg	1012.54	x
P246	Nphe-Nphe-Npcb-Npcb	1098.53	x
P247	Nphe-Nphe-Npcb-Mys	1045.59	x
P248	Nphe-Nphe-Npcb-Nphe	1064.57	x
P249	Nphe-Nphe-Mys-Nprg	959.60	x
P250	Nphe-Nphe-Mys-Npcb	1045.59	x
P251	Nphe-Nphe-Mys-Mys	992.65	x
P252	Nphe-Nphe-Mys-Nphe	1011.63	x
P253	Nphe-Nphe-Nphe-Nprg	978.58	x
P254	Nphe-Nphe-Nphe-Npcb	1064.57	0
P255	Nphe-Nphe-Nphe-Mys	1011.63	x

MTT assay – Library 1

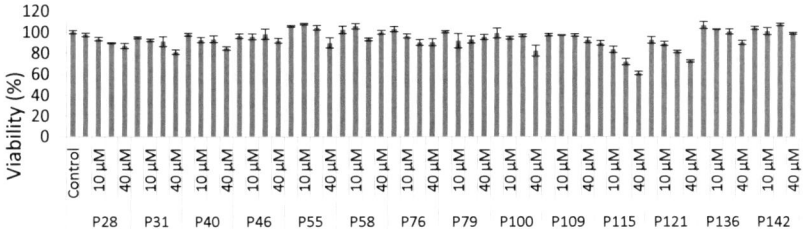

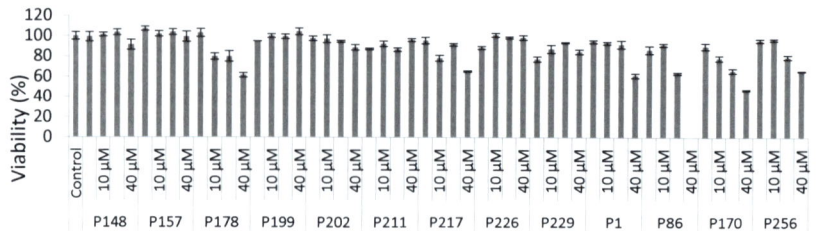

Figure 79: Cytotoxicity of representative peptoids in library 1 in HeLa cells. Each peptoid contains Nlys, Nprg, Nphe and Npcb once without doubling one side chain. Peptoids were tested in 5, 10, 20 and 40 µM concentration for 72 h and viability was determined using the MTT assay.

Spheroids

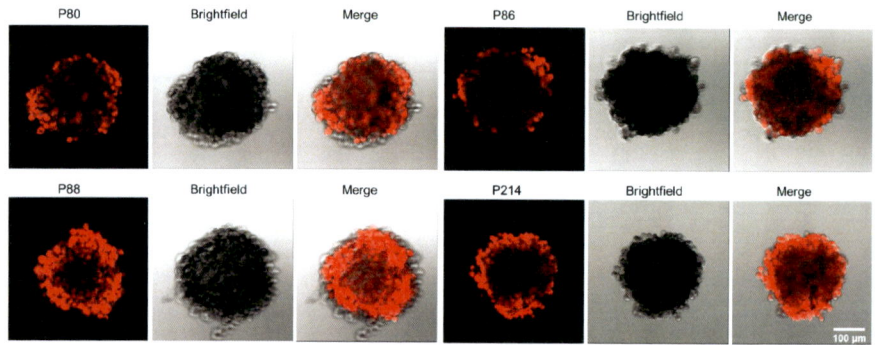

Figure 80: Potential anticancer peptoids P80, P86, P88 and P214 incubated on SK-MEL 28 spheroids (4 x 10³ cells) for 24 h. Spheroids were washed with DPBS and subsequently analyzed by fluorescent confocal microscopy using a Leica TCS-SPE microscope (objective: 20x/0.70 DRY UV). 1. Peptoid (Ex.: 561 nm, Em.: 593-696 nm), 2. Brightfield, 3. Merge

Reactive oxygen species- peptoids without dye

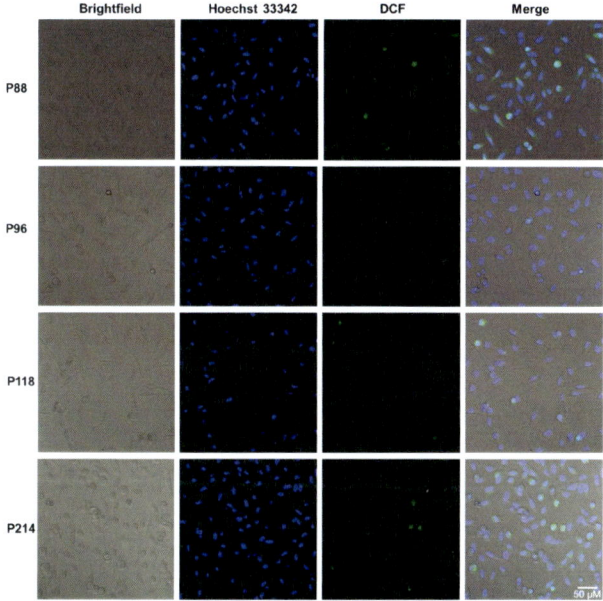

Figure 81: Detection of intracellular reactive oxygen species in HeLa cells induced by peptoids without fluorescent dye. 1.5 x 10^4 cells were treated with 30 µM of the respective peptoid for 6 h at 37 °C. Reactive oxygen species were detected by incubation with H_2DCFDA. For co-staining of nuclei cells were treated with Hoechst 33342 (2 µg/ml). Intracellular generation of ROS was investigated by fluorescence confocal microscopy (Leica TCS-SPE, Objective: 20x/0.70 DRY UV). 1. Brightfield 2. Hoechst 33342, Ex.: 364nm, Em.: 417-468 nm, 3. DCF Ex.: 488 nm, Em.: 510-520 nm 4. Merge

Reactive oxygen species

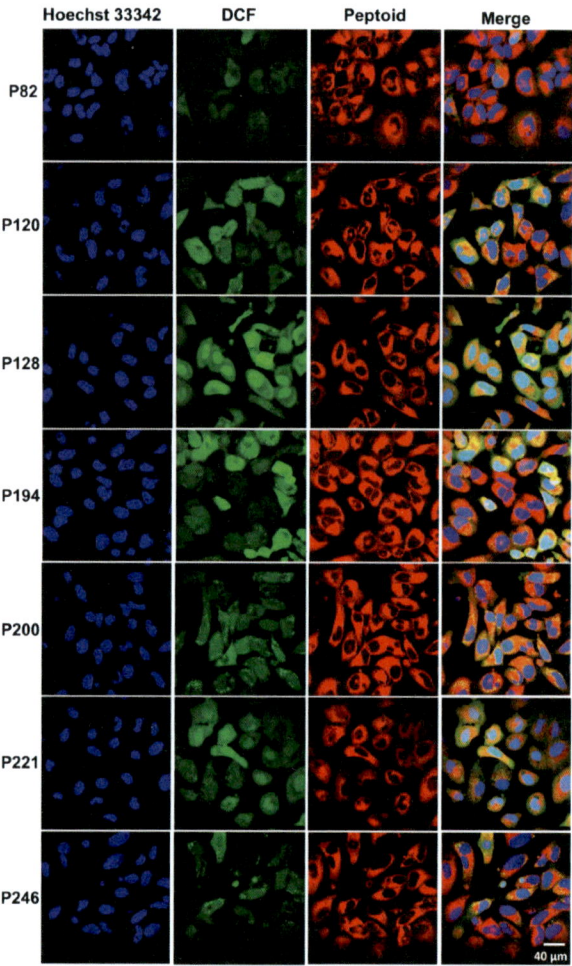

Figure 82: Detection of intracellular reactive oxygen species in HeLa cells induced by peptoids. 1.5 x 10⁴ cells were treated with 30 µM of the respective peptoid for 6 h at 37 °C. Reactive oxygen species were detected by incubation with H₂DCFDA. For co-staining of nuclei cells were treated with Hoechst 33342 (2 µg/ml). Intracellular generation of ROS was investigated by fluorescence confocal microscopy (Leica TCS-SPE, Objective: ACS APO 63x/1.30 OIL). 1. Hoechst 33342, Ex.: 364nm, Em.: 417-468 nm, 2. DCF Ex.: 488 nm, Em.: 510-520 nm 3. Peptoid, Ex.: 561 nm, Em.: 593-696 nm 4. Merge

Peptoid categories – zebrafish screenings

Eye and yolk sac

Table 18: Peptoid numbers and sequences accumulating in eye and yolk sac of 4 dpf zebrafish larvae incubated for 2 h with 50 µM peptoid solution.

P4	Nprg-Nprg-Nprg-Nphe	P104	Npcb-Mys-Npcb-Nphe	P193	Nphe-Nprg-Nprg-Nprg
P13	Nprg-Nprg-Nphe-Nprg	P122	Npcb-Nphe-Mys-Npcb	P208	Nphe-Nprg-Nphe-Nphe
P49	Nprg-Nphe-Nprg-Nprg	P127	Npcb-Nphe-Nphe-Mys	P218	Nphe-Npcb-Mys-Npcb
P95	Npcb-Npcb-Nphe-Mys	P152	Mys-Npcb-Npcb-Nphe		
P110	Npcb-Mys-Nphe-Npcb	P160	Mys-Npcb-Nphe-Nphe		

Digestive system

Table 19: Peptoid numbers and sequences accumulating in the digestive system of 4 dpf zebrafish larvae incubated for 2 h with 50 µM peptoid solution.

P1	Nprg-Nprg-Nprg-Nprg	P50	Nprg-Nphe-Nprg-Npcb	P140	Mys-Nprg-Mys-Nphe
P5	Nprg-Nprg-Npcb-Nprg	P52	Nprg-Nphe-Nprg-Nphe	P141	Mys-Nprg-Nphe-Nprg
P9	Nprg-Nprg-Mys-Nprg	P65	Npcb-Nprg-Nprg-Nprg	P144	Mys-Nprg-Nphe-Nphe
P10	Nprg-Nprg-Mys-Npcb	P93	Npcb-Npcb-Nphe-Nprg	P153	Mys-Npcb-Mys-Nprg
P21	Nprg-Npcb-Npcb-Nprg	P109	Npcb-Mys-Nphe-Nprg	P177	Mys-Nphe-Nprg-Nprg
P29	Nprg-Npcb-Nphe-Nprg	P125	Npcb-Nphe-Nphe-Nprg	P180	Mys-Nphe-Nprg-Nphe
P36	Nprg-Mys-Nprg-Nphe	P129	Mys-Nprg-Nprg-Nprg	P189	Mys-Nphe-Nphe-Nprg
P41	Nprg-Mys-Mys-Nprg	P132	Mys-Nprg-Nprg-Nphe	P192	Mys-Nphe-Nphe-Nphe
P44	Nprg-Mys-Mys-Nphe	P138	Mys-Nprg-Mys-Npcb	P240	Nphe-Mys-Nphe-Nphe

Apoptotic cells

Table 20: Peptoid numbers and sequences inducing apoptotic cells in 4 dpf zebrafish larvae incubated for 2 h with 50 µM peptoid solution.

P26	Nprg-Npcb-Mys-Npcb	P166	Mys-Mys-Npcb-Npcb	P220	Nphe-Npcb-Mys-Nphe
P90	Npcb-Npcb-Mys-Npcb	P181	Mys-Nphe-Npcb-Nprg	P226	Nphe-Mys-Nprg-Npcb
P103	Npcb-Mys-Npcb-Mys	P182	Mys-Nphe-Npcb-Npcb	P230	Nphe-Mys-Npcb-Nphe
P106	Npcb-Mys-Mys-Npcb	P190	Mys-Nphe-Nphe-Npcb	P231	Nphe-Mys-Npcb-Mys
P119	Npcb-Nphe-Npcb-Mys	P109	Npcb-Mys-Nphe-Nprg	P232	Nphe-Mys-Npcb-Nphe
P149	Mys-Npcb-Npcb-Nprg	P184	Mys-Nphe-Npcb-Nphe	P250	Nphe-Nphe-Mys-Npcb
P151	Mys-Npcb-Npcb-Mys	P211	Nphe-Npcb-Nprg-Mys	P252	Nphe-Nphe-Mys-Nphe
P154	Mys-Npcb-Mys-Npcb	P215	Nphe-Npcb-Npcb-Mys		
P155	Mys-Npcb-Mys-Mys	P219	Nphe-Npcb-Mys-Mys		

Caudal vein

Table 21: Peptoid numbers and sequences accumulating in the caudal vein of 4 dpf zebrafish larvae incubated for 2 h with 50 µM peptoid solution.

P6	Nprg-Nprg-Npcb-Npcb	P112	Npcb-Mys-Nphe-Nphe	P170	Mys-Mys-Mys-Npcb
P81	Npcb-Npcb-Nprg-Nprg	P115	Npcb-Nphe-Nprg-Mys	P171	Mys-Mys-Mys-Mys
P92	Npcb-Npcb-Mys-Nphe	P123	Npcb-Nphe-Mys-Mys	P183	Mys-Nphe-Npcb-Mys
P98	Npcb-Mys-Nprg-Npcb	P124	Npcb-Nphe-Mys-Nphe	P167	Mys-Mys-Npcb-Mys
P99	Npcb-Mys-Nprg-Mys	P147	Mys-Npcb-Nprg-Mys	P174	Mys-Mys-Nphe-Npcb
P101	Npcb-Mys-Npcb-Nprg	P158	Mys-Npcb-Nphe-Npcb	P234	Nphe-Mys-Mys-Npcb
P111	Npcb-Mys-Nphe-Mys	P159	Mys-Npcb-Nphe-Mys		

High toxicity

Table 22: Peptoid numbers and sequences leading to strong toxicity for 4 dpf zebrafish larvae incubated for 2 h with 50 µM peptoid solution.

P18	Nprg-Npcb-Nprg-Npcb	P84	Npcb-Npcb-Nprg-Nphe	P150	Mys-Npcb-Npcb-Npcb
P22	Nprg-Npcb-Npcb-Npcb	P85	Npcb-Npcb-Npcb-Nprg	P184	Mys-Nphe-Npcb-Nphe
P23	Nprg-Npcb-Npcb-Mys	P86	Npcb-Npcb-Npcb-Npcb	P198	Nphe-Nprg-Npcb-Npcb
P24	Nprg-Npcb-Npcb-Nphe	P87	Npcb-Npcb-Npcb-Mys	P200	Nphe-Nprg-Npcb-Nphe
P30	Nprg-Npcb-Nphe-Npcb	P88	Npcb-Npcb-Npcb-Nphe	P214	Nphe-Npcb-Npcb-Npcb
P36	Nprg-Mys-Nprg-Nphe	P93	Npcb-Npcb-Nphe-Nprg	P216	Nphe-Npcb-Npcb-Nphe
P40	Nprg-Mys-Npcb-Nphe	P96	Npcb-Npcb-Nphe-Nphe	P219	Nphe-Npcb-Mys-Mys
P54	Nprg-Nphe-Npcb-Npcb	P102	Npcb-Mys-Npcb-Npcb	P221	Nphe-Npcb-Nphe-Nprg
P69	Npcb-Nprg-Npcb-Nprg	P114	Npcb-Nphe-Nprg-Npcb	P223	Nphe-Npcb-Nphe-Mys
P70	Npcb-Nprg-Npcb-Npcb	P118	Npcb-Nphe-Npcb-Npcb	P224	Nphe-Npcb-Nphe-Nphe
P71	Npcb-Nprg-Npcb-Mys	P120	Npcb-Nphe-Npcb-Nphe	P232	Nphe-Mys-Npcb-Nphe
P72	Npcb-Nprg-Npcb-Nphe	P126	Npcb-Nphe-Nphe-Npcb	P239	Nphe-Mys-Nphe-Mys
P79	Npcb-Nprg-Nphe-Mys	P128	Npcb-Nphe-Nphe-Nphe	P247	Nphe-Nphe-Npcb-Mys
P80	Npcb-Nprg-Nphe-Nphe	P146	Mys-Npcb-Nprg-Npcb		
P82	Npcb-Npcb-Nprg-Npcb				

Lateral line

Table 23: Peptoid numbers and sequences accumulating in the lateral line (neuromast cells) of 4 dpf zebrafish larvae incubated for 2 h with 50 µM peptoid solution.

P11	Nprg-Nprg-Mys-Mys	P76	Npcb-Nprg-Mys-Nphe	P188	Mys-Nphe-Mys-Nphe
P12	Nprg-Nprg-Mys-Nphe	P83	Npcb-Npcb-Nprg-Mys	P191	Mys-Nphe-Nphe-Mys
P15	Nprg-Nprg-Nphe-Mys	P100	Npcb-Mys-Nprg-Nphe	P195	Nphe-Nprg-Nprg-Mys
P25	Nprg-Npcb-Mys-Nprg	P105	Npcb-Mys-Mys-Nprg	P199	Nphe-Nprg-Npcb-Mys
P27	Nprg-Npcb-Mys-Mys	P121	Npcb-Nphe-Mys-Nprg	P201	Nphe-Nprg-Mys-Nprg

P28	Nprg-Npcb-Mys-Nphe	P130	Mys-Nprg-Nprg-Npcb	P202	Nphe-Nprg-Mys-Npcb
P30	Nprg-Npcb-Nphe-Npcb	P131	Mys-Nprg-Nprg-Mys	P203	Nphe-Nprg-Mys-Mys
P31	Nprg-Npcb-Nphe-Mys	P133	Mys-Nprg-Npcb-Nprg	P204	Nphe-Nprg-Mys-Nphe
P33	Nprg-Mys-Nprg-Nprg	P135	Mys-Nprg-Npcb-Mys	P207	Nphe-Nprg-Nphe-Mys
P35	Nprg-Mys-Nprg-Mys	P137	Mys-Nprg-Mys-Nprg	P217	Nphe-Npcb-Mys-Nprg
P39	Nprg-Mys-Npcb-Mys	P139	Mys-Nprg-Mys-Mys	P225	Nphe-Mys-Nprg-Nprg
P42	Nprg-Mys-Mys-Npcb	P161	Mys-Mys-Nprg-Nprg	P227	Nphe-Mys-Nprg-Mys
P45	Nprg-Mys-Nphe-Nprg	P163	Mys-Mys-Nprg-Mys	P228	Nphe-Mys-Nprg-Nphe
P46	Nprg-Mys-Nphe-Npcb	P165	Mys-Mys-Npcb-Nprg	P229	Nphe-Mys-Npcb-Nprg
P47	Nprg-Mys-Nphe-Mys	P168	Mys-Mys-Npcb-Nphe	P233	Nphe-Mys-Mys-Nprg
P51	Nprg-Nphe-Nprg-Mys	P169	Mys-Mys-Mys-Nprg	P235	Nphe-Mys-Mys-Mys
P55	Nprg-Nphe-Npcb-Mys	P172	Mys-Mys-Mys-Nphe	P236	Nphe-Mys-Mys-Nphe
P63	Nprg-Nphe-Nphe-Mys	P175	Mys-Mys-Nphe-Mys	P237	Nphe-Mys-Nphe-Nprg
P67	Npcb-Nprg-Nprg-Mys	P176	Mys-Mys-Nphe-Nphe	P243	Nphe-Nphe-Nprg-Mys
P73	Npcb-Nprg-Mys-Nprg	P178	Mys-Nphe-Nprg-Npcb	P249	Nphe-Nphe-Mys-Nprg
P74	Npcb-Nprg-Mys-Npcb	P186	Mys-Nphe-Mys-Npcb	P255	Nphe-Nphe-Nphe-Mys
P75	Npcb-Nprg-Mys-Mys	P187	Mys-Nphe-Mys-Mys		

Olfactory system

Table 24: Peptoid numbers and sequences accumulating in the olfactory system of 4 dpf zebrafish larvae incubated for 2 h with 50 µM peptoid solution.

P8	Nprg-Nprg-Npcb-Nphe	P242	Nphe-Nphe-Nprg-Npcb	P304	Npbf -Npob-Nphe-Npob
P14	Nprg-Nprg-Nphe-Npcb	P244	Nphe-Nphe-Nprg-Nphe	P305	Npbf -Npob-Npbf-Nphe
P17	Nprg-Npcb-Nprg-Nprg	P245	Nphe-Nphe-Npcb-Nprg	P307	Npbf-Npob-Npbf-Npob
P20	Nprg-Npcb-Nprg-Nphe	P246	Nphe-Nphe-Npcb-Npcb	P308	Npbf-Npob-Npob-Nphe
P56	Nprg-Nphe-Npcb-Nphe	P263	Nphe-Nphe-Npob-Nphe	P309	Npbf-Npob-Npob-Npbf
P61	Nprg-Nphe-Nphe-Nprg	P268	Nphe-Npbf-Nphe-Npob	P310	Npbf-Npob-Npob-Npob
P62	Nprg-Nphe-Nphe-Npcb	P269	Nphe-Npbf-Npbf-Nphe	P317	Npob -Nphe-Npob-Nphe
P64	Nprg-Nphe-Nphe-Nphe	P274	Nphe-Npbf-Npob-Npob	P319	Npob -Nphe-Npob-Npob
P66	Npcb-Nprg-Nprg-Npcb	P275	Nphe-Npob-Nphe-Nphe	P328	Npob -Npbf-Npob-Npob
P68	Npcb-Nprg-Nprg-Nphe	P276	Nphe-Npob-Nphe-Npbf	P329	Npob -Npob-Nphe-Nphe
P77	Npcb-Nprg-Nphe-Nprg	P277	Nphe-Npob-Nphe-Npob	P331	Npob -Npob-Nphe-Npob
P94	Npcb-Npcb-Nphe-Npcb	P278	Nphe-Npob-Npbf-Nphe	P334	Npob-Npob-Npbf-Npob
P116	Npcb-Nphe-Nprg-Nphe	P280	Nphe-Npob-Npbf-Npob	P335	Npob-Npob-Npob-Nphe
P117	Npcb-Nphe-Npcb-Nprg	P281	Nphe-Npob-Npob-Nphe	P337	Npob-Npob-Npob-Npob
P125	Npcb-Nphe-Nphe-Nprg	P282	Nphe-Npob-Npob-Npbf		
P194	Nphe-Nprg-Nprg-Npcb	P286	Npbf -Nphe-Nphe-Npob		
P205	Nphe-Nprg-Nphe-Nprg	P290	Npbf -Nphe-Npob-Nphe		
P206	Nphe-Nprg-Nphe-Npcb	P292	Npbf -Nphe-Npob-Npob		

P210	Nphe-Npcb-Nprg-Npcb	**P298**	Npbf -Npbf-Npbf-Npob
P212	Nphe-Npcb-Nprg-Nphe	**P300**	Npbf -Npbf-Npob-Npbf
P241	Nphe-Nphe-Nprg-Nprg	**P301**	Npbf -Npbf-Npob-Npob

Olfactory system and lateral line

Table 25: Peptoid numbers and sequences accumulating in the olfactory system and the lateral line (neuromast cells) of 4 dpf zebrafish larvae incubated for 2 h with 50 µM peptoid solution.

P257	Nphe-Nphe-Nphe-Nphe	**P273**	Nphe-Npbf-Npob-Npbf	**P293**	Npbf -Npbf-Nphe-Nphe
P264	Nphe-Nphe-Npob-Npbf	**P276**	Nphe-Npob-Nphe-Npbf	**P294**	Npbf -Npbf-Nphe-Npbf
P266	Nphe-Npbf-Nphe-Nphe	**P280**	Nphe-Npob-Npbf-Npob	**P296**	Npbf -Npbf-Npbf-Nphe
P268	Nphe-Npbf-Nphe-Npob	**P285**	Npbf -Nphe-Nphe-Npfb	**P313**	Npob-Nphe-Nphe-Npob
P271	Nphe-Npbf-Npbf-Npob	**P287**	Npbf -Nphe-Npbf-Nphe	**P318**	Npob-Nphe-Npob-Npbf
P272	Nphe-Npbf-Npob-Nphe	**P291**	Npbf -Nphe-Npob-Npbf	**P320**	Npob -Npbf-Nphe-Nphe

Toxic or unspecific

Table 26: Peptoid numbers and sequences leading to toxicity or unspecific accumulation in 4 dpf zebrafish larvae incubated for 2 h with 50 µM peptoid solution.

P259	Nphe-Nphe-Nphe-Npob	**P303**	Npbf -Npob-Nphe-Npbf	**P320**	Npob -Npbf-Nphe-Nphe
P261	Nphe-Nphe-Npbf-Npbf	**P306**	Npbf-Npob-Npbf-Npbf	**P323**	Npob-Npbf-Npbf-Nphe
P267	Nphe-Npbf-Nphe-Npbf	**P311**	Npob-Nphe-Nphe-Nphe	**P325**	Npob-Npbf-Npbf-Npob
P269	Nphe-Npbf-Npbf-Nphe	**P312**	Npob-Nphe-Nphe-Npfb	**P327**	Npob-Npbf-Npob-Npbf
P284	Npbf-Nphe-Nphe-Nphe	**P314**	Npob -Nphe-Npbf-Nphe	**P330**	Npob-Npob-Nphe-Npbf
P288	Npbf -Nphe-Npbf-Npbf	**P315**	Npob-Nphe-Npbf-Npbf-		
P289	Npbf -Nphe-Npbf-Npob	**P316**	Npob-Nphe-Npbf-Npob		

9. Curriculum Vitae

Personal Information

Name	Ilona Diana Majella Wehl
Date of birth and City	15/11/1990 in Berlin
Nationality	German

Education

03/2016 – 02/2019	PhD student, Karlsruhe Institute of Technology, Karlsruhe
	Group: Prof. Dr. Ute Schepers
	Member of the Research Training Group 2039 – Molecular Architecture for fluorescent cell imaging, funded by DFG (Deutsche Forschungsgemeinschaft).
08/2017 – 12/2017	Visiting Scholar at Lawrence Berkeley National Laboratory, Berkeley, CA, USA
	Group: Dr. Ron Zuckermann
	"Synthesis of peptoids and binding of fluorescent bacteria to functionalized peptoid nanosheets"
04/2014 – 02/2016	M.Sc., Chemical Biology, Karlsruhe Institute of Technology, Karlsruhe
	Major subjects: Organic chemistry and biochemistry
	Thesis: High-throughput microscopy and combinatorial synthesis of peptoids for the identification of organ specific transporters"
10/2010 – 02/2014	B.Sc., Chemical Biology, Karlsruhe Institute of Technology, Karlsruhe
	Thesis: Enzymatic synthesis of short chain hyaluronic acid oligomers with defined length using a microfluidic device for cell- free enzymatic synthesis
09/2001 – 06/2010	Elly-Heuss-Knapp-Gymnasium, Heilbronn, Abitur

Professional Experience

03/2015	Scientific assistant, Karlsruher Institute of Technology, Karlsruhe
	Group: Prof. Dr. Stefan Bräse

Skills and workshops

Languages	German (native), English (fluent), French (intermediate), Spanish (basic)
Workshops	HPLC Troubleshooting (12/2016)

GxP course: GLP, GMP, GCLP (04/2016)

Communication and Presentation Skills Exemplified for Entrepreneurship (03/2016)

Leadership training (BASF, "Führungstraining" (04/2017)

Visual Communication (11/2018)

International Zebrafish and Medaka course (10/2015)

Scholarships and prices

Scholarships Research Travel Grant of Karlsruhe House of Young Scientists (KHYS)

Funding Funded member of Research Training Group 2039 by DFG

Prices 02/2016: Master with distinction

Publications

[1] A.A. Popova, D. Marcato, R. Peravali, I. Wehl, U. Schepers, P. Levkin, Fish-Microarray: A Miniaturized Platform for Single-Embryo High-Throughput Screenings. Advanced Functional Materials, 2018. **28**(3): p. 1703486.

[2] A. B. Braun, I. Wehl, D. Kölmel, U. Schepers, S. Bräse, Novel polyfluorinated cyanine dyes for the selective NIR-staining of mitochondria, **2019**, accepted by Chem. Eur. J.

Posters in Conferences

I. Wehl, F. Rönicke, M. Reischl, R. Mikut, U. Schepers, High-throughput screenings and combinatorial synthesis of peptoids to identify organ specific transporter molecules, 10th Peptoid Summit, 2017, Berkeley, USA

10. Acknowledgement

Firstly, I would like to express my very great appreciation to my adviser Prof. Dr. Ute Schepers for giving me the opportunity to work on this suspenseful topic, helpful discussions and advices, motivation and encouragement. I am grateful for the scientific support but also the freedom in shaping this project.

Thank you also to Prof. Dr. Hans-Achim Wagenknecht for being the co-referee of this thesis.

Furthermore, I would like to thank the Karlsruhe House of Young Scientists for providing me with a research travel grant and Dr. Ron Zuckermann for the possibility to expand my work at the Molecular Foundry (Lawrence Berkeley National Laboratory).

Thanks also to Prof. Dr. Ralf Mikut and Dr. Markus Reischl for the pleasant collaboration. I would also like to thank Prof. Dr. Stefan Bräse and his working group for hosting me in his labs. Moreover, my thanks go to Dipl.-Ing. Ravindra Peravali, for lots of help with the screening microscopes and Dr. Sepand Rastegar for knowledge and discussions about zebrafish.

I would also like to express particular thanks to Nadine Borel and everyone in the KIT fish facility for the support and help, especially during the last month of my PhD.

Besides, I would like to thank Dr. Franziska Rönicke and all other collaboration partners for our joint work, enthusiastic discussions and many ideas to improve this work.

A very special gratitude goes out to my working group (Anna Meschkov, Bettina Fleck, Christoph Grün, Dr. Eva Zittel, Rebecca Pfister, Sonja Haase, Tobias Göckler, Dr. Vanessa Kappings and Xenia Kempter) who are not only colleagues, but also friends. Thank you for cheering up my day, always having an open ear for problems, motivation and funny evenings and trips. Special thanks also for proof reading this work.

I am thankful that I had the possibility to perform this work within the Research Training Group 2039 (funded by the Deutsche Forschungsgemeinschaft), giving me the opportunity to improve this work in collaboration with other members, as well as interesting and helpful seminars and retreats.

Furthermore, I would like to extend my thanks my bachelor students Dominik Feser, Jessica Stevens and Xenia Kempter for their contribution to this work.

Last but not least, a very special thanks to my parents Majella Wehl and Prof. Dr. Wolfgang Wehl, my sister Ines Galm and her family and Daniel Holub. Without your support, help, encouragement and motivation during the last years this wouldn´t be possible.